Biomechanics in Orthodontics: Principles and Practice

Biomechanics in Orthodontics

PRINCIPLES AND PRACTICE

Ram S. Nanda, BDS, DDS, MS, PhD

Professor Emeritus
Department of Orthodontics
College of Dentistry
University of Oklahoma
Oklahoma City, Oklahoma

Yahya S. Tosun, DDS, PhD

Private Practice
Dubai, United Arab Emirates

Former Professor
Department of Orthodontics
University of Aegea
İzmir, Turkey

Quintessence Publishing Co, Inc

Chicago, Berlin, Tokyo, London, Paris, Milan, Barcelona,
Istanbul, Moscow, New Delhi, Prague, São Paulo, and Warsaw

Library of Congress Cataloguing-in-Publication Data

Nanda, Ram S., 1927-
 Biomechanics in orthodontics : principles and practice / Ram S. Nanda, Yahya Tosun.
   p. ; cm.
 Includes bibliographical references and index.
 ISBN 978-0-86715-505-1
 1. Orthodontic appliances. 2. Biomechanics. I. Tosun, Yahya. II. Title.
 [DNLM: 1. Biomechanics. 2. Orthodontic Appliances. 3. Malocclusion—therapy. 4. Orthodontic Appliance Design. WU 426 N176b 2010]
 RK527.N366 2010
 617.6'430284—dc22
                            2010013547

© 2010 Quintessence Publishing Co, Inc

Quintessence Publishing Co Inc
4350 Chandler Drive
Hanover Park, IL 60133
www.quintpub.com

All rights reserved. This book or any part thereof may not be reproduced, stored in a retrieval system, or transmitted in any form or by any means, electronic, mechanical, photocopying, or otherwise, without prior written permission of the publisher.

Editor: Lisa C. Bywaters
Design: Gina Ruffolo
Production: Sue Robinson

Printed in China

Contents

Preface vii

1 Physical Principles *1*

2 Application of Orthodontic Force *17*

3 Analysis of Two-Tooth Mechanics *55*

4 Frictional and Frictionless Systems *71*

5 Anchorage Control *83*

6 Correction of Vertical Discrepancies *99*

7 Correction of Transverse Discrepancies *125*

8 Correction of Anteroposterior Discrepancies *133*

9 Space Closure *145*

Glossary 154

Index 156

Preface

Once comprehensive diagnosis and treatment planning have set the stage for initiating treatment procedures, appliance design and systems have to be developed to achieve treatment goals. Correct application of the principles of biomechanics assists in the selection of efficient and expedient appliance systems.

Over the last three decades, there has been an explosion in the development of technology related to orthodontics. New materials and designs for brackets, bonding, and wires have combined to create a nearly infinite number of possibilities in orthodontic appliance design. As these new materials are brought together in the configuration of orthodontic appliances, it is necessary to understand and apply the principles of biomechanics for a successful and efficient treatment outcome. Lack of proper understanding may not only set up inefficient force systems but also cause collateral damage to the tissues. The path to successful treatment is through good knowledge of biomechanics.

This book is written with the purpose of introducing a student of orthodontics to the evolving technology, material properties, and mechanical principles involved in designing orthodontic appliances.

Physical Principles

CHAPTER 1

Movement of teeth in orthodontic treatment requires application of forces and periodontal tissue response to these forces. Force mechanics are governed by physical principles, such as the laws of Newton and Hooke. This chapter presents the basic definitions, concepts, and applicable mechanical principles of tooth movement, laying the groundwork for subsequent chapters.

Newton's Laws

Isaac Newton's (1642–1727) three laws of motion, which analyze the relations between the effective forces on objects and their movements, are all applicable to clinical orthodontics.

The law of inertia

The law of inertia analyzes the static balance of objects. Every body in a state of rest or uniform motion in a straight line will continue in the same state unless it is compelled to change by the forces applied to it.

The law of acceleration

The law of acceleration states that the change in motion is proportional to the motive force that is applied. Acceleration occurs in the direction of the straight line in which the force is applied: $a = F/m$, where a = acceleration, F = force, and m = mass.

The law of action and reaction

The reaction of two objects toward each other is always equal and in an opposite direction. Therefore, to every action there is always an equal and opposite reaction.

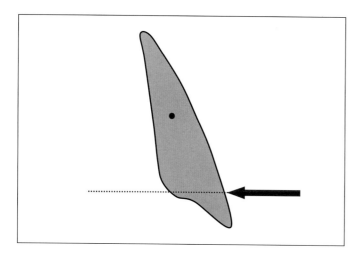

Fig 1-1 Force is a vector. The force applied to the incisor is signified by the length of the arrow, and the point of application is on the crown. Its line of action is horizontal, and its direction is from anterior to posterior.

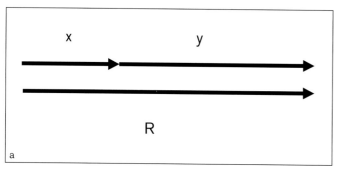

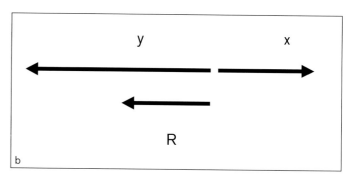

Fig 1-2 The resultant (R) of forces (x and y) on the same line of action and direction is $R = x + y$ (a) and the same line of action but in different directions is $R = x + (-y)$ (b).

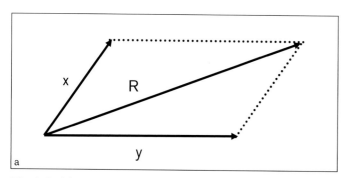

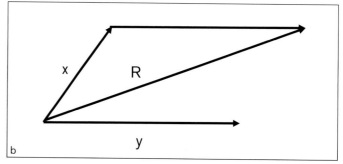

Fig 1-3 (a) The resultant (R) of the x and y vectors that have the same point of origin is the diagonal of the parallelogram with these vectors used as the sides. (b) R can also be obtained by drawing a vector parallel to vector y and extending from the tip of vector x, then drawing a line joining its tip to the origin of vector x.

Vectors

When any two points in space are joined, a line of action is created between these points. When there is movement from one of these points toward the other, a direction is defined. The magnitude of this force is called a vector, it is shown by the length of an arrow, and its point of application is shown with a point. For example, in Fig 1-1, the line of action of the force vector, which is applied by the labial arch of a removable appliance on the labial surface of the crown of the incisor, is horizontal. The direction is backward (ie, from anterior to posterior), and its amount is signified by the length of the arrow.

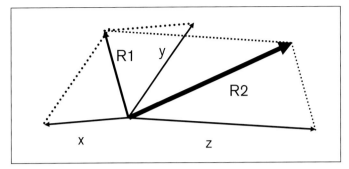

Fig 1-4 To find the sum of multiple vectors having the same point of origin, first draw the resultant (R1) of vectors x and y, thus defining the resultant (R2) of the z and R1 vectors (ie, $x + y = R1$; $z + R1 = R2$).

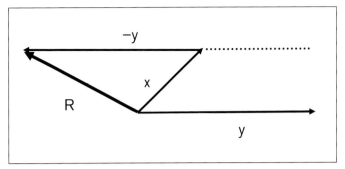

Fig 1-5 The difference between x and y vectors having the same point of origin can be obtained by drawing a vector (–y) starting from the tip of vector x that runs parallel to the y vector but in the opposite direction; then the tip of vector –y is joined to the point of origin of vectors x and y.

Addition of vectors

Vectors are defined in a coordinate system. The use of two coordinate axes can be sufficient for vectors on the same plane.

In Fig 1-2a, the resultant (R) of the vectors of different forces (x and y), which are on the same line of action and in the same direction, equals the algebraic sum of these two vectors $(x + y)$. The resultant of two vectors on the same line of action but in opposite directions can be calculated as $(x + [-y])$ (Fig 1-2b).

The resultant of two vectors that have a common point of origin is the diagonal of a parallelogram whose sides are the two vectors (Fig 1-3a). The resultant of the same vectors can also be obtained by joining the tip of a vector parallel to vector y drawn from the tip of vector x to the point of origin of vector x (Fig 1-3b).

Sum of multiple vectors

The sum of multiple vectors is calculated in the same system as the calculation of two vectors. Therefore, the third vector is added to the resultant of the first two vectors, and so on (Fig 1-4).

Subtraction of two vectors

To define the difference between two vectors, a new vector (–y) is drawn in the opposite direction from the tip of vector x and parallel to vector y, and the point of origin of vector x is joined to the tip of vector –y (Fig

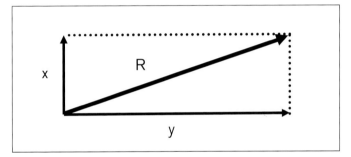

Fig 1-6 The separation of a resultant vector into components on an x- and y-axis coordinate system.

1-5). Thus, the resultant (R) is from the point of origin of vectors x and y toward the tip of vector –y.

Separating a vector into components

To separate a resultant vector (R) into components, two parallel lines are drawn from the point of origin of that vector toward the components that are searched. By drawing parallels from the R vector's tip toward these lines, a parallelogram is obtained. The sum of the two components obtained by this method is exactly equal to vector R.

The separation of a resultant vector into components is generally (at the elementary level) realized on x and y reference axes for ease of presentation and trigonometric calculations (Fig 1-6). In fact, for complicated calculations, vectors can be separated into unnumbered directions. Therefore, the x-axis is generally accepted

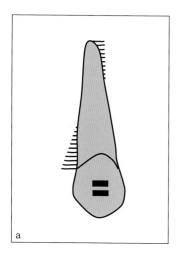
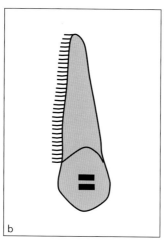

Fig 1-7 Distribution of force on the bone and root surfaces in tipping (a) and translatory (b) movements. During tipping, the possibility of indirect bone resorption is high because the forces are concentrated in small areas. Therefore, the forces must be kept as low as possible.

as the horizontal axis, and the *y*-axis is accepted as the vertical axis. Thus, the component x of vector R can be defined as horizontal and the component y as vertical.

Force

Force is the effect that causes an object in space to change its place or its shape. In orthodontics, the force is measured in grams, ounces, or Newtons. Force is a vector having the characteristics of line of action, direction, magnitude, and point of application. In the application of orthodontic forces, some factors such as distribution and duration are also important. During tipping of a tooth, force is concentrated at the alveolar crest on one side and at the apex on the other (Fig 1-7a). During translation, however, the force is evenly distributed onto the bone and root surfaces (Fig 1-7b).

Forces according to their duration

Constancy of force
Clinically, optimal force is the amount of force resulting in the fastest tooth movement without damage to the periodontal tissues or discomfort to the patient. To achieve an optimum biologic response in the periodontal tissues, light, continuous force is important.[1] Figure 1-8 compares the amount of loss of force occurring over time on the force levels of two coil springs of high and low load/deflection rates.[2]

Continuous forces A continuous force can be obtained by using wires with low load/deflection rate and high working range. In the leveling phase, where there is considerable variation in level between teeth, it is advantageous to use these wires to control anchorage and maintain longer intervals between appointments. Continuous force depreciates slowly, but it never diminishes to zero within two activation periods (clinically, this period is usually 1 month); thus, constant and controlled tooth movement results[3] (Fig 1-9a). For example, the force applied by nickel titanium (NiTi) open coil springs is a continuous force.

Interrupted forces Interrupted forces are reduced to zero shortly after they have been applied. If the initial force is relatively light, the tooth will move a small amount by direct resorption and then will remain in that position until the appliance is reactivated. After the application of interrupted forces, the surrounding tissues undergo a repair process until the second activation takes place[3] (Fig 1-9b). The best example of an active element that applies interrupted force is the rapid expansion screw.

Intermittent forces During intermittent force application, the force is reduced to zero when the patient removes the appliance[3] (Fig 1-9c). When it is placed back into the mouth, it continues from its previous level, reducing slowly. Intermittent forces are applied by extraoral appliances.[3]

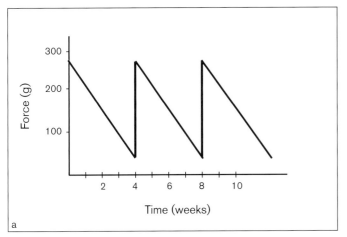 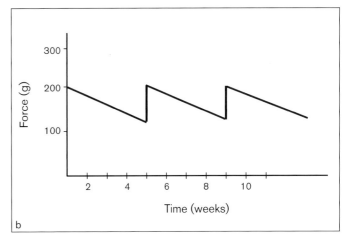

Fig 1-8 The loss of force, with time, in springs with high (a) and low (b) load/deflection rates. In the same period (4 weeks), the loss of force in the spring with a high load/deflection rate was approximately 225 g, compared to only 75 g in the spring with a low load/deflection rate. (Reprinted from Gjessing[2] with permission.)

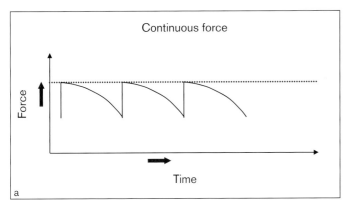 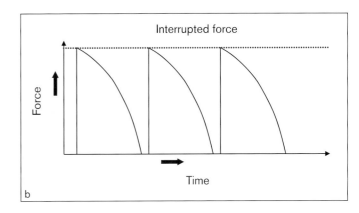

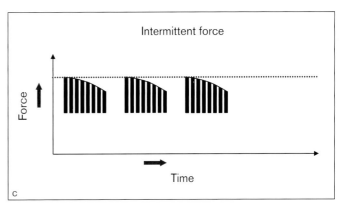

Fig 1-9 The effects of continuous (a), interrupted (b), and intermittent (c) forces on the periodontal tissues. (Reprinted from Proffit[3] with permission.)

Center of Resistance

The point where the line of action of the resultant force vector intersects the long axis of the tooth, causing translation of the tooth, is defined as the *center of resistance*. Theoretically, the center of resistance of a tooth is located on its root, but the location has been extensively investigated. Studies show that the center of resistance

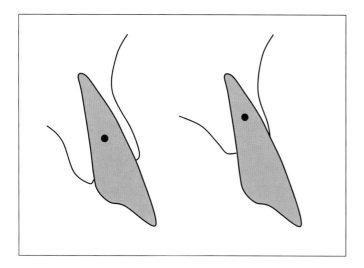

Fig 1-10 The center of resistance moves apically in response to a loss of alveolar bone or periodontal attachment.

of single-rooted teeth is on the long axis of the root, approximately 24% to 35% of the distance from the alveolar crest.[4–10]

The center of resistance is sometimes confused with the center of mass. The center of mass is a balance point of a free object in space under the effect of gravity. A tooth, however, is a restrained object within the periodontal and bony structures surrounded by muscle forces. Therefore, the center of resistance must be considered a balance point of restrained objects.

The center of resistance is unique for every tooth; the location of this point depends on the number of roots, the level of the alveolar bone crest, and the length and morphology of the roots. Therefore, the center of resistance sometimes changes with root resorption or loss of alveolar support because of periodontal disease (Fig 1-10). For example, in the case of loss of alveolar support, this point moves apically.[11]

Center of Rotation

The center of rotation is the point around which the tooth rotates. The location of this point is dependent on the force system applied to the tooth, that is, the moment-to-force (M/F) ratio. When a couple of force is applied on the tooth, this point is superimposed on the center of resistance (ie, the tooth rotates around its center of resistance). In translation it becomes infinite, meaning there is no rotation. This subject is explained in greater detail in the M/F ratio section later in the chapter.

Moment

Moment is the tendency for a force to produce rotation or tipping of a tooth. It is determined by multiplying the magnitude of the force (F) by the perpendicular distance (d) from the center of resistance to the line of action of this force: $M = F \times d$ (Fig 1-11). In orthodontic practice, it is usually measured in gram-millimeters, or g-mm, which means grams $\times$ millimeters. Forces passing through the center of resistance do not produce a moment, because the distance to the center of resistance is zero. Hence, the tooth does not rotate; it translates (Fig 1-12). Because the moment depends on the magnitude of the force and the perpendicular distance to the center of resistance, it is possible to obtain the same rotational effect by doubling the distance and reducing the magnitude of the force by half, or vice versa. Even when the force is not excessive but the distance from the center of resistance to the line of action is significant, the periodontal tissues may be adversely affected because of the large moment.

Couple

A couple is a system having two parallel forces of equal magnitude acting in opposite directions. Every point of a body to which a couple is applied is under a rotational effect in the same direction and magnitude. No matter where the couple is applied, the object rotates about its center of resistance—that is, the center of resistance and

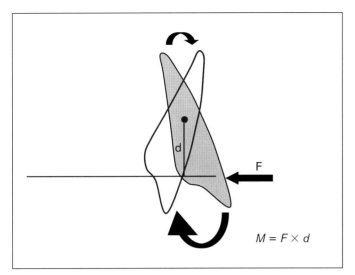

Fig 1-11 The line of action of any force (F) not passing through the center of resistance creates a moment (M), which is a rotational or tipping effect on the tooth. According to the formula $M = F \times d$, a moment is proportional to the magnitude of force and the distance (d) perpendicular from its line of action to the center of resistance.

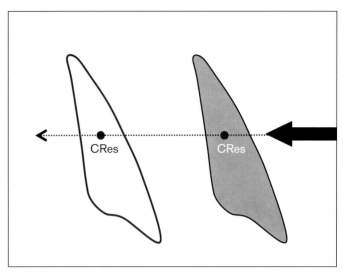

Fig 1-12 A force having a line of action passing through the center of resistance (CRes) causes translation of the tooth. During this movement, the center of resistance moves along the line of action of the force.

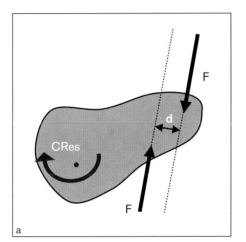
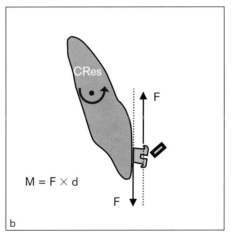
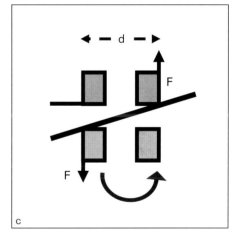

Fig 1-13 A couple causes an object to rotate around its center of resistance regardless of the point of application, thereby superimposing the center of resistance (CRes) and the center of rotation (a). Two examples of fixed appliances in which the couple is applied are torque in the third order (b) and antitip in the second order (c). In calculating the moment (M) of a couple, it is sufficient to multiply the magnitude of one of the forces (F) by the perpendicular distance (d) between the lines of action of these forces.

the center of rotation superimpose[12] (Fig 1-13). For example, a torque (third-order couple) applied to an incisor bracket causes tipping of the tooth about its center of resistance. This phenomenon is explained in detail in the equivalent force systems section later in the chapter. The calculation of the moment of a couple can be performed by multiplying the magnitude of one of the forces by the perpendicular distance between the lines of action.

Transmissibility of a Force Along Its Line of Action

Forces can be transmitted along their line of action without any change in their physical sense. Provided that the line of action is the same, any force acting on a tooth would be equally effective if it were applied by pushing distally with an open coil or pulling distally with a chain

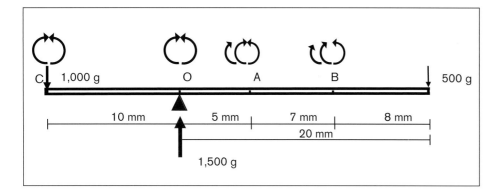

Fig 1-14 On a scale in a state of static equilibrium, the moments around all the points must be in balance. The algebraic total of the moments on the O, A, B, and C points on the bar that tend to rotate the system clockwise and counterclockwise must equal zero.

elastic. The principle of transmissibility states that the external effect of a force acting on a tooth is independent of where the force is applied along its line of action.[13]

Static Equilibrium and the Analysis of Free Objects

The rules of static equilibrium are applied similarly for every object or mechanical system and for every part of that object or system. Therefore, to make it easier to understand the forces applied on a mechanical system, it is sufficient to analyze only a part of the system as a free object. For instance, to define all the forces applied on a dental arch, it may be sufficient to analyze the relations between only 2 teeth instead of analyzing all 14 teeth. Of course, the forces applied in this system of 2 teeth must be in balance. Briefly, the analysis of a free object is the study of an isolated part of a system or an object in a state of static equilibrium, enabling us to get an idea about the whole system.

Statics deals with the state of an object in equilibrium under the influence of forces. The main law of statics is Newton's first law, which implies that if a body is at a state of rest or in stable motion in a certain direction, the resultant of the forces acting on this body is zero. In other words, static equilibrium implies that at any point within a body, the algebraic sum of all the effective forces on the body in space should be zero ($\Sigma Fx = 0$; $\Sigma Fy = 0$; $\Sigma Fz = 0$). For the body to be in balance in the sense of rotation, the algebraic sum of all the moments effective on it must also be zero ($\Sigma Mx = 0$; $\Sigma My = 0$; $\Sigma Mz = 0$). The sum of the moments acting on a body around any point of a body in static equilibrium state is zero. An example is shown numerically in Fig 1-14. Orthodontically, understanding this law is very important because it is fundamental to every clinical application.

The book in Fig 1-15 is in a state of rest. The factor that enables this book to remain still is the fact that a force (A), the weight of the book, has an equal and opposite force (N), the table, working against it. Because the system is statically balanced, there is balance between the forces acting on it. For an object to be in a state of static equilibrium, the foremost condition is that there must be no motion in the system.

Tooth Movement

Tipping

Controlled and uncontrolled tipping
Tipping, in practice, is the easiest type of tooth movement. When a single force is applied to a bracket on a round wire, the tooth tips about its center of rotation, located in the middle of the root, close to its center of

Fig 1-15 The book on the table is in a state of rest. For the book to remain statically balanced, a force (N) of the same magnitude as the weight of the book (A) but in the opposite direction must act on the book.

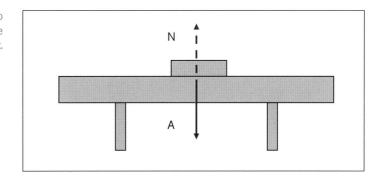

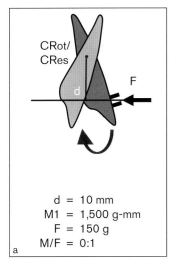

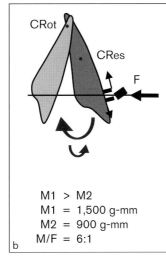

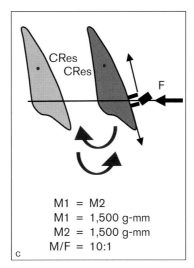

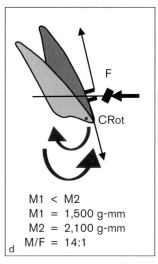

a
d = 10 mm
M1 = 1,500 g-mm
F = 150 g
M/F = 0:1

b
M1 > M2
M1 = 1,500 g-mm
M2 = 900 g-mm
M/F = 6:1

c
M1 = M2
M1 = 1,500 g-mm
M2 = 1,500 g-mm
M/F = 10:1

d
M1 < M2
M1 = 1,500 g-mm
M2 = 2,100 g-mm
M/F = 14:1

Fig 1-16 A change in the M/F ratio applied to the tooth will cause its center of rotation to change its position. In the uncontrolled tipping movement (M/F ratio = 0:1), the center of rotation (CRot) is located very close to the center of resistance (CRes) (a), whereas in controlled tipping (M/F ratio = 6:1), it is located near the apex (b). In *translation* (M/F ratio = 10:1), the center of rotation is infinite (ie, there is no rotation) (c). In root movement (M/F ratio = 14:1), this point is located near the crown (d). d, distance; F, force.

resistance. This single force causes movements of the crown and apex in opposite directions. This movement, caused by the moment of force (M1), is called *uncontrolled tipping*[13] (Fig 1-16a), and it is usually clinically undesirable. In this movement, the M/F ratio can vary from approximately 0:1 to 1:5 (see the M/F ratio section later in the chapter).

If a light, counterclockwise moment (M2; torque) is added to the system with a rectangular wire while the single distal force is still being applied, the tooth tips distally in what is called *controlled tipping*, which is clinically desirable. In this movement, the center of rotation moves apically, and the tooth tips around a circle of a greater radius. In controlled tipping, the M/F ratio is from approximately 6:1 to 9:1 (Fig 1-16b).

When the counterclockwise moment (M2; torque) is increased to equal the moment of force (M1), the moments neutralize each other, and there is no rotation in the system. In this case, the center of rotation no longer exists (it is infinite) and the tooth undergoes translation, or bodily movement (Fig 1-16c). In translation, the M/F ratio is approximately 10:1 to 12:1. Clinically, translation is a desirable movement, but it is hard to achieve and maintain. If the counterclockwise moment (M2; torque) is increased even more to an M/F ratio of approximately 14:1, then the moment becomes greater than M1, and the tooth undergoes root movement. In root movement, the center of rotation is located at the crown (Fig 1-16d).

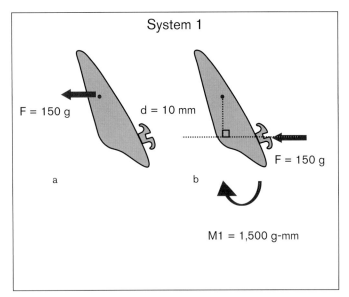 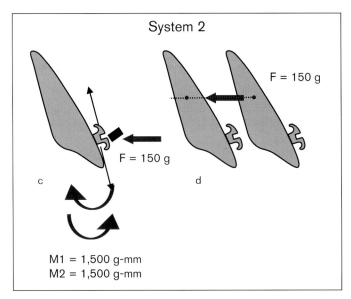

Fig 1-17 *(a)* Because translating an incisor by applying a force that passes directly through the center of resistance is not practical, this movement can be achieved by setting up an equivalent force system on the crown that gives the same result. *(b)* When a force (F) of 150 g is applied to the crown, a clockwise moment (M1) of 1,500 g-mm occurs. *(c)* If this moment is balanced with an equal and opposite moment (M2), only a net force of 150 g remains in the system. *(d)* Even though it is applied to the crown, this force causes translation as if it were being applied to the center of resistance. d, distance.

Translation (bodily movement)

Theoretically, translation of a body is the movement of any straight line on that body, without changing the angle with respect to a fixed reference frame (see Fig 1-12). During translation, all the points on the body move the same distance, and they therefore have the same velocity.

Rotation

Rotation of a body is the movement of any straight line on that body by a change in the angle with respect to a fixed reference frame. If the body rotates about its center of resistance, it is called *pure rotation*.

Equivalent Force Systems

As stated earlier, it is possible to slide forces along their lines of action without changing their physical sense. However, it is impossible to slide them parallel to their lines of action, because by changing the location of the line of action of a force, the distance to the center of resistance also changes. Therefore, the type of tooth movement changes (see Figs 1-11 and 1-12). The equivalent force system principle, illustrated in Fig 1-17, states that the same type of translatory movement obtained by the force (F) passing through the center of resistance can be achieved by application of force at the bracket on the crown. Forces passing through the center of resistance cause translation. Clinically, it is not always practical to apply forces through the center of resistance of a tooth (System 1) because of anatomical and biomechanical limitations. Therefore, this force system must be replaced with another system (System 2) that is applied to the crown. Forces applied to the crown usually do not pass through the center of resistance. They cause rotation (or tipping) of the tooth because of the clockwise moment (M1). To obtain translation, this moment must be balanced by an equal and opposite moment (M2) (ie, $M1 = M2 = 1{,}500$ g-mm). In the example, M2 can be obtained with palatal root torque applied to the incisor bracket (Fig 1-17c). As a result, the moments neutralize each other, and the single force of 150 g remains on the system that translates the tooth.

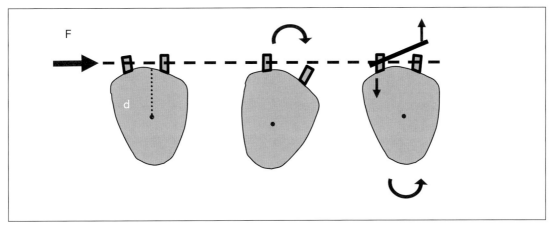

Fig 1-18 The type of tooth movement in the sagittal direction in Fig 1-17 is also valid in the transverse direction. The only difference is that a couple (antirotation) is applied in the first order. d, distance; F, force.

Moment-to-Force Ratio

Understanding the M/F ratio concept is vital to the clinician in controlling tooth movements (see Fig 1-16).[14] Clinically, the M/F ratio determines the types of movement or the location of the center of rotation.[10,11,13] According to the formula, $M/F = F \times d/F = d$, the M/F ratio equals the distance (d). As the distance between the center of resistance and the line of action increases, the M/F ratio also increases. If we take the example in Fig 1-16a, a distal force of 150 g is applied to the incisor bracket. Because the force does not pass through the center of resistance (d = 10 mm), and an opposite moment (M2) is not applied to the bracket, the M/F ratio of this system is 0:1 (there is no moment of couple). The tooth tips distally about the center of rotation, located on the root, close to the center of resistance. This is uncontrolled tipping in which the crown moves distally and the apex mesially; this is the easiest type of tooth movement to achieve from a clinical standpoint. In removable appliances, springs or screws cause uncontrolled tipping because there is only a single force acting on the tooth, and there is no attachment on the tooth for an opposite moment (see Fig 1-11). Likewise, a similar type of movement is observed in the Begg technique in which the teeth move on round wires.

If a counterclockwise moment (M2) of 900 g-mm is applied on the bracket with palatal root torque, the M/F ratio becomes 6:1 (see Fig 1-16b). In this case, the center of rotation moves apically, so the tooth moves as a pendulum around its apex (or a point close to it). This is a case of controlled tipping.

If the M2 moment is increased to 1,500 g-mm, then the M/F ratio becomes 10:1. The moments balance each other, and only 150 g of single force remains on the system, causing the tooth to translate. In this case, the center of rotation of the tooth is infinite (see Fig 1-16c). If the magnitude of the M2 moment is increased even more, up to 2,100 g-mm, the M/F ratio becomes 14:1. In this case, the center of rotation moves to the crown. This is root movement (see Fig 1-16d).

Everything explained above is also valid in the transverse plane. A canine moving distally with a segmented arch acting on the bracket at a point away from the center of resistance rotates distolingually. This rotation can be controlled with an antirotation bend. In this plane, the M/F ratio is also equal to the distance between center of resistance and the line of action of the force (Fig 1-18).

In the second order, as a result of distal force, the canine tips distally. This moment is balanced with an antitip bend (counterclockwise moment). The magnitude of this moment depends on the amount of the bend and the width of the bracket. Note that for the same amount of couple, forces applied on the wings of a narrow bracket are higher compared with the wide bracket in the second order. This is basically because of the difference in distance between parallel forces of the couple. The greater the distance, the less the force and vice versa. Because the distance on the narrow bracket is less than

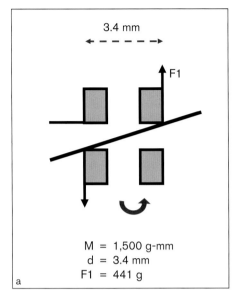 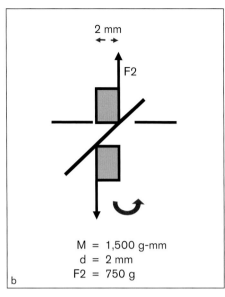 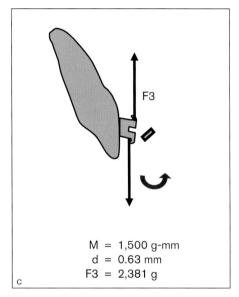

Fig 1-19 For the same moment (M), the force applied to the wings of the brackets decreases as the width increases. Because the distance (d) on the wide bracket is larger (3.4 mm) *(a)* than that of the narrow bracket (2 mm) *(b)*, the force is low. In torque *(c)*, the magnitude of force on the bracket wings is high because the distance (0.63 mm) is very small.

that of the wide bracket, the magnitude of forces is higher. For example, assuming the width of the brackets (d) is 3.4 mm, the amount of force applied on the bracket wings can be calculated (Figs 1-19a and 1-19b) as

$$M = F \times d$$
$$1{,}500 = F \times 3.4$$
$$F = 1{,}500/3.4 = 441 \text{ g}$$

If a narrow bracket (2 mm) is used, the amount of force effective on the bracket wing is 750 g. During application of a third-order couple (torque), because the interbracket distance is very small, the magnitude of force on the bracket wings is that much more (Fig 1-19c). This is one of the main reasons ceramic bracket wings break when torqued.

M/F ratios of teeth with loss of alveolar support

The center of resistance of a tooth depends on the length, number, and morphology of the roots and the level of alveolar bone support. In root resorption, the root shortens, causing the center of resistance to move occlusally, but with loss of alveolar bone support, the center of resistance moves apically (Fig 1-20; see also Fig 1-10). This is particularly important in the treatment of adult patients, who often have periodontal problems. As the distance (d) between bracket and center of resistance increases, the M/F ratio also increases. To obtain a higher M/F ratio, two options can be considered:

- Place the bracket gingivally. If this is done, the bracket base may no longer adapt well to the tooth surface.[15] Furthermore, it will be more difficult to insert a straight wire for leveling. An adaptive step-up bend may be needed on the archwire, but this might affect the precision of alignment.
- Increase the moment, decrease the force, or a combination of both.[15] Clinically, the moment applied to the bracket is predictable only if a segmented archwire is used. The moment generated by an antitip or torque bend cannot be measured accurately. It is therefore difficult to accurately adjust the M/F ratio by changing the moment. Adjusting the amount of force according to the type of movement obtained seems to be more practical.

Braun et al[15] have stated that the M/F ratio depends basically on the location of the center of resistance, and they have determined coefficients of moments and

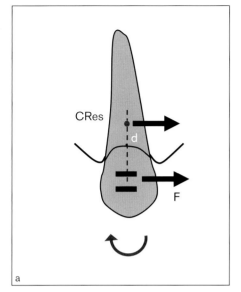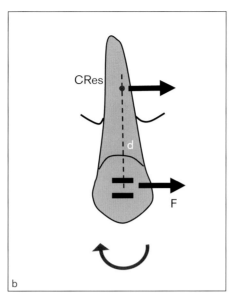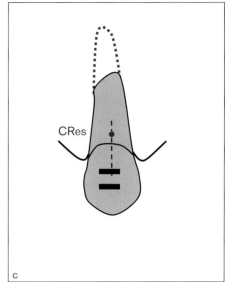

Fig 1-20 In the case of alveolar bone loss (a), the center of resistance (CRes) moves apically, thus the distance (d) increases (b). For translation, the M/F ratio should also be increased. Clinically, it is better to reduce the force (F) to control tooth movement. This is particularly important in adults who may have alveolar bone loss owing to periodontal disease. In the case of root resorption (c), the center of resistance moves occlusally. (Reprinted with permission from Braun et al.[15])

Table 1-1 Moment and force values that can be applied to a tooth with alveolar bone loss*

Loss of alveolar support for offset increase (mm)	Moment multiplying factor to compensate for offset increase	Force multiplying factor to compensate
0	–	–
1	1.06	0.94
2	1.13	0.89
3	1.19	0.84
4	1.25	0.80
5	1.32	0.76

*Reprinted with permission from Braun et al.[15]

forces to be used in cases with loss of alveolar bone support (Table 1-1).

Clinically, determining the position of the center of resistance and the exact value of the M/F ratio to be applied and keeping it stable throughout the movement is fairly difficult. Tanne et al[5] have stated that because of a very small change in the M/F ratio and the exponential relationship between the center of rotation of the tooth and this ratio, the center of rotation can change drastically. The amount of force is key to controlling the M/F ratio (ie, tooth movement). If undesired tipping occurs as a result of an overactivated loop, one should let the wire work until root movement has been accomplished.

Effect of loop configuration on M/F ratio

The objective of a loop is to decrease the load/deflection rate of the wire (ie, increase the elasticity); thus, to apply force in a wider range within physiologic limits. Because of spring characteristics, loops are commonly used in space closure with differential mechanics. During space closure, it is important to obtain controlled tooth movement in both anterior and posterior segments. Uncon-

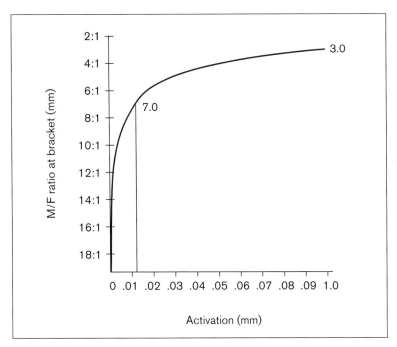

Fig 1-21 The M/F ratios of a 6-mm-high vertical loop with a 20-degree gable bend. During deactivation of the loop, the M/F ratio shows a relatively stable curve throughout a distance of 0.9 mm, then increases dramatically up to 19:1 at the last 0.1 mm. (Reprinted with permission from Burstone.[17])

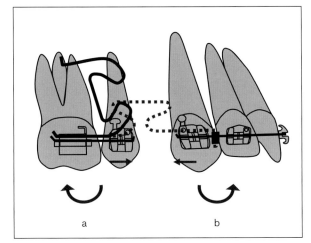

Fig 1-22 The activation (a) and neutral (b) positions of a T-loop.

trolled tipping is usually undesirable because side effects such as anchorage loss and root resorption may occur in uprighting. For example, with the Begg appliance, extraction spaces are usually closed in a relatively short time by uncontrolled tipping of the anterior teeth, using a combination of round wires and Class II elastics. Uprighting excessively tipped incisors, however, takes longer and requires a high level of anchorage support.[16]

The configuration of the loop has a significant effect on the M/F ratio. Studies have shown that the M/F ratio generated by a vertical loop is approximately 2:1.[17] Increasing the length of the loop can increase the M/F ratio to 4:1,[18] but increasing the height of the loop is not practical because it may cause irritation and discomfort.

Figure 1-21 shows the changes in the M/F ratio of a vertical loop 6 mm high and having 20 degrees of antitip. Note that the M/F ratio generated by a 1-mm activation is less than 3:1, which produces uncontrolled tipping. The M/F ratio increases as the loop deactivates. When it has deactivated to 0.1 mm, the M/F ratio approaches 7:1, which approximates controlled tipping. The M/F ratio increases to approximately 20:1 when the deactivation goes from 0.1 to 0 mm.

It is evident that 0.1 mm of activation is clinically significant. A minor error in activation could change the location of the center of rotation dramatically. In clinical terms, with a 1.0-mm activation, the tooth undergoes uncontrolled tipping during the first 0.7 mm of deactivation (M/F ratio, 5:1). The center of rotation shifts to a point between the center of resistance and the apex. Thereafter, the tooth sustains controlled tipping from 0.3 to 0.12 mm. The M/F ratio becomes 7:1 and the center of rotation moves to a point between the apex and infinity. As the loop deactivates from 0.12 mm to 0.03 mm, the M/F ratio becomes 10:1, the center of rotation disappears (goes to infinity), and the tooth translates. At the final deactivation between 0.03 and 0 mm,

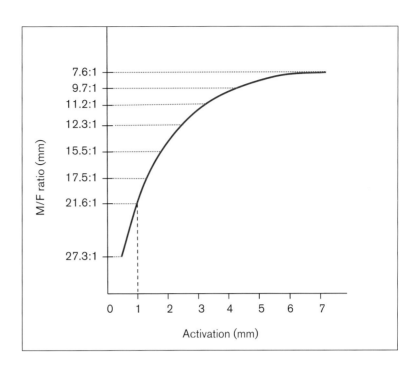

Fig 1-23 M/F ratios achieved by 7-mm activation of a T-loop on a 0.016 × 0.022–inch TMA wire. The M/F ratio generated by the T-loop activated 1 mm is approximately 7 times more (21.6:1) than that of a vertical loop (3:1). At full activation (7 mm), a T-loop generates an M/F ratio of 7.6:1, which produces controlled tipping. (Based on data from Manhartsberger et al.[18])

the M/F ratio rises to 20:1, and the center of rotation moves occlusally to a point close to the crown. In this case, the tooth undergoes root movement.[17]

The amount of force produced by vertical loops per unit of activation can be relatively high. For example, the force produced by a Bull loop of 0.018 × 0.025–inch stainless steel (SS) wire delivers approximately 500 g.[19] Such a large force produces uncontrolled tipping, and it may cause root resorption. If we need only 100 g of force, then the activation of the loop would be only 0.2 mm, which is clinically impractical. After activation, the legs of a vertical loop close rapidly, which initially produces uncontrolled tipping. If one lets the wire work, the tooth uprights slowly with root movement. The duration of uprighting depends on the amount of tipping and the M/F ratio. The more the tooth tips, the longer the uprighting will take.

Clinically, a good space-closing loop should generate an M/F ratio high enough to produce controlled tipping at maximum activation. The M/F ratio increases progressively as the loop deactivates and produces root movement. As mentioned earlier, increasing the length or adding a helix to the vertical loop decreases its load/deflection rate but has little effect on the M/F ratio.[16] Putting more wire at the top of the loop to increase the M/F ratio is recommended.[17,18] There are two reasons for incorporating more wire in a loop:

- To increase the M/F ratio (if placed gingivally)
- To decrease the load/deflection rate

A T-loop made from β-titanium (β-titanium alloy, or β-Ti; also titanium-molybdenum alloy, or TMA) wire is capable of generating a higher M/F ratio at a larger activation than a vertical loop. To do so, a T-loop should be "preactivated" before bracket engagement (Fig 1-22). The preactivation (gable) bend can sometimes reach 180 degrees from the horizontal, according to the anchorage needs of the case. A 0.017 × 0.025–inch TMA T-loop preactivated 180 degrees and activated 7 mm horizontally delivers approximately 350 g.[13] A 0.016 × 0.022–inch TMA T-loop with the same activation generates 243 g,[18] while a 0.017 × 0.025–inch and an 0.018–inch TMA composite T-loop deliver 333 g.[17] To obtain 150 g for canine distalization, a T-loop on a 0.016 × 0.022–inch archwire requires only 4 mm of activation.

Figure 1-23 shows an M/F ratio generated by a T-loop on a 0.016 × 0.022–inch TMA wire.[18] An M/F ratio at 7 mm activation is 7.6:1, which produces controlled tipping. The average force per unit of activation (1 mm) of the T-loop is 34.5 g, which is relatively low. An error of 1 mm in activation of the loop applies 34.5 g more force to the periodontal tissues.

The M/F ratio increases as the loop deactivates. When it has deactivated to 2.7 mm, the M/F ratio approaches 12:1, which approximates root movement. The M/F ratio increases to approximately 27:1 when the deactivation goes from 2.7 to 0.5 mm.

Conclusion

The laws of mechanics governing forces and motion in the context of orthodontics concern material properties (eg, the stress, strain, stiffness, springiness, and elastic limit of wires). Concepts such as moments and couples, center of resistance and center of rotation, and the moment-to-force ratio are likewise essential to understand in order to control tooth movements. This basic knowledge of the physical principles behind orthodontics allows the practitioner to design appliances and plan treatment that will provide optimal results.

References

1. Reitan K. Biomechanical principles and reactions. In: Graber TM (ed). Orthodontics: Current Principles and Techniques. St Louis: Mosby, 1985:118.
2. Gjessing P. Biomechanical design and clinical evaluation of a new canine-retraction spring. Am J Orthod 1985;87:353–362.
3. Proffit WR. Contemporary Orthodontics. St Louis: Mosby, 1986:235,237,238.
4. Burstone CJ, Pryputniewicz RJ. Holographic determination of centers of rotation produced by orthodontic forces. Am J Orthod 1980;77:396–409.
5. Tanne K, Koenig HA, Burstone CJ. Movement to force ratios and the center of rotation. Am J Orthod Dentofacial Orthop 1988;94:426–431.
6. Tanne K, Nagataki T, Inoue Y, Sakuda M, Burstone CJ. Patterns of initial tooth displacements associated with various root lengths and alveolar bone heights. Am J Orthod Dentofacial Orthop 1991;100:66–71.
7. Tanne K, Sakuda M, Burstone CJ. Three-dimensional finite element analysis for stress in the periodontal tissue by orthodontic forces. Am J Orthod Dentofacial Orthop 1987;92:499–505.
8. Vanden Bulcke MM, Burstone CJ, Sachdeva RJ, Dermaut LR. Location of centers of resistance for anterior teeth during retraction using the laser reflection technique. Am J Orthod Dentofacial Orthop 1987;91:375–384.
9. Pedersen E, Andersen K, Gjessing P. Electronic determination of centres of rotation produced by orthodontic force systems. Eur J Orthod 1990;12:272–280.
10. Smith RJ, Burstone CJ. Mechanics of tooth movement. Am J Orthod 1984;85:294–307.
11. Nanda R, Kuhlberg A. Principles of biomechanics. In: Nanda R (ed). Biomechanics in Clinical Orthodontics. Philadelphia: Saunders, 1996:3.
12. Kusy RP, Tulloch JF. Analysis of moment/force ratios in the mechanics of tooth movement. Am J Orthod Dentofacial Orthop 1986;90:127–131.
13. Marcotte MR. Biomechanics in Orthodontics. Philadelphia: Decker, 1990:63.
14. Burstone CJ. Application of bioengineering to clinical orthodontics. In: Graber TM (ed). Orthodontics: Current Principles and Techniques. St Louis: Mosby, 1985:198–199.
15. Braun S, Winzler J, Johnson BE. An analysis of orthodontic force systems applied to the dentition with diminished alveolar support. Eur J Orthod 1993;15:73–77.
16. Staggers JA, Germane N. Clinical considerations in the use of retraction mechanics. J Clin Orthod 1991;25:364–369.
17. Burstone CJ. The segmented arch approach to space closure. Am J Orthod 1982;82:361–378.
18. Manhartsberger C, Morton JY, Burstone CJ. Space closure in adult patients using the segmented arch technique. Angle Orthod 1989;59:205–210.
19. Ulgen M. Treatment Principles in Orthodontics, ed 3. Ankara, Turkey: Ankara University, 1990.

Application of Orthodontic Force

The treatment of malocclusion of teeth requires fabrication of orthodontic appliances to activate force systems. Various elements constitute an appliance. Starting with brackets, banded or bonded to the teeth, the actual forces are applied from the archwires, coil springs, and elastic modules. There are many variables involved in the structural and material properties of each component. It is important to understand these variables for construction of an efficient and tissue-friendly appliance.

The energy sources used for orthodontic tooth movement are forces originating from the elasticity of wires and elastics. A light continuous force is required for optimum orthodontic tooth movement. When the force level is maintained throughout treatment, there is a smooth progression of tooth movement resulting from direct bone resorption,[1] and undesirable side effects, such as loss of anchorage or damage to the periodontal tissues, are avoided. As the tooth moves, the magnitude of force is gradually reduced owing to the structural characteristics of wires and elastics. In orthodontic practice, it is desirable to maintain a force at optimum level throughout tooth movement. To achieve this goal, the use of superelastic wires is preferred because they remain active for a long time and deliver forces at a physiologic level.

Because wires and elastics are the main sources of force application in orthodontics, it is important to know their physical properties.

Physical Properties of Materials Used in Orthodontics

Materials are composed of molecules and atoms. The distance between the grains (particles) and the connective force between them defines the hardness of the materials. When force is applied to a material, the distance between the atoms changes, depending on the nature of the force, which is called *stress*. The material becomes strained by the movement of the force–the stress–until it changes in size and becomes deformed.

When the stress on the material is a pulling force, the distance between the atoms increases and the material expands. This process is called *tension*. When the stress is a pushing force, the distance between the atoms is reduced and the size of the material decreases. This process is called *compression*.

2 | Application of Orthodontic Force

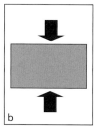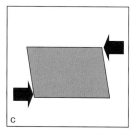

Fig 2-1 The forms of stress applied on a material: tension *(a)*; compression *(b)*; shear *(c)*.

When a pair of equal and opposite forces, called a *couple*, are applied to the material from different planes, a third kind of deformation occurs.[2] This process is called *shear*. Tension, compression, and shear are the forms of stress that can be applied to a material (Fig 2-1).

Rubber stretches with the application of a pulling force, whereas a steel rod will not be noticeably deformed by the same amount of force. For the same amount of stress, the strain on materials varies from one to the other. Deformation of the steel rod cannot be seen with the naked eye, for instance, but there may be a change of size that can be measured microscopically.

Elastic behavior of materials

A material under stress absorbs energy from the force and returns it when the stress is removed. Materials that can return the absorbed energy completely, regaining their original size when the stress is removed, are called *elastic* materials. Materials unable to return to their original size are called *plastic* materials. Coil springs are elastic, for example, whereas dead soft ligature wires are plastic. Between these extremes, viscoelastic materials can show elastic and plastic behaviors at the same time. Examples of these materials are human skin, muscles, veins, nerves, and fibers.

The concept of elasticity has a significant bearing on orthodontics because the elasticity of materials is the most important source of force used in appliance systems. An orthodontic appliance consists of *active* and *passive* units. The foremost elements that constitute the active units are sources of force such as wires, coil springs, and elastics. Activation of an orthodontic appliance or its active elements can be estimated by measuring the magnitude of the force with a dynamometer or simply by observing the deformation of the active elements.

When a piece of wire is bent up to a certain point and let go, it returns to its original position. But if the force is increased excessively, and the wire is bent beyond a certain point, it cannot return to its original position. This results in plastic deformation of the wire. Figure 2-2 shows the strain of a wire onto which stress is applied. As the stress (force) is increased, the deflection of the wire increases proportionally; this constitutes the linear portion of the curve. This linear portion rises to the elastic limit of the wire, meaning that the wire will return to its original position upon removal of the stress.

The principle that applies to this section of the curve is called the *Hook law*, which states that "the strain is proportional to the stress applied on the material up to the elastic limit." The slope of the linear portion of the curve, identified as the *stress/strain rate*, gives the modulus of elasticity, also known as the *Young modulus*. This value (represented by the letter E in Fig 2-2) expresses the stiffness, or load/deflection rate (springiness), of the material. Stiffness and springiness are reciprocal properties; that is, stiffness is proportional and springiness is inversely proportional to the modulus of elasticity.[1–6]

When the elastic limit is exceeded, the wire cannot return to its original shape upon removal of the stress because plastic deformation has occurred. When additional stress is applied, permanent deformation of the wire increases and it reaches the ultimate strength limit. Beyond this point, the molecular structure of the wire breaks down and it snaps at the failure point.

Elastic, or polymeric, materials with amorphous structure, such as rubber, show a different stress-strain curve than do metals, such as the wire in the previous example, which have crystal structure (Fig 2-3).[3] When stress is first applied to an elastic, there is a linear change as there is with metal, but this linear distance is much shorter than that of metal. In elastics, the elastic limit of the material is reached sooner than in metals, so permanent deformation occurs more readily. As with metals, when the ultimate strength limit of an elastic is exceeded, the material fails.

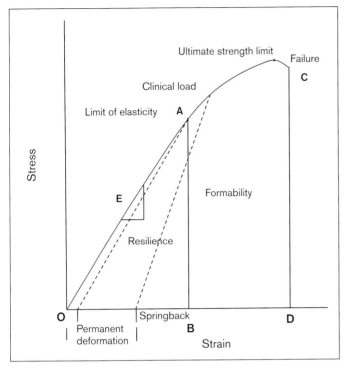

Fig 2-2 Stress-strain graph of a wire under stress. According to the Hook law, stress and strain are proportional up to the elastic limit. The limit of elasticity at point A will permit the wire to return to its original position at point O. The wire on which force is loaded up to this point and let go will return to its original position. The angle of the slope gives the modulus of elasticity (ie, the Young modulus, E) of the wire. When the elastic limit is exceeded, permanent deformation of the wire occurs. When the ultimate strength limit is reached, the molecular structure of the wire is broken down and the wire fails. The area between the points O, A, and B gives the modulus of resilience; the area between the points A, B, D, and C gives the formability; and the total area under the slope from the starting point to the failure point gives the modulus of toughness. (Redrawn from Proffit[1] with permission.)

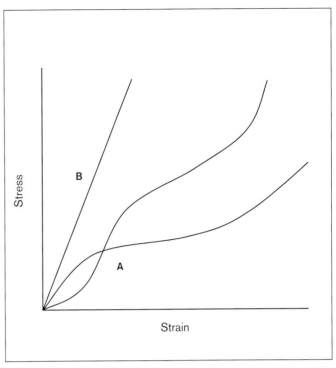

Fig 2-3 Strain-stress graph of elastic (A) and ceramic (B) materials. (Reprinted from Nikolai[3] with permission.)

The B slope shown in Fig 2-3 belongs to a ceramic material that does not have any elasticity, and it shows a straight line until the failure point. Materials with these properties are said to be *brittle*.[3]

In a stress-strain diagram (see Fig 2-2), the area under the slope up to the elastic limit gives the modulus of resilience of the material. Resilience is the amount of energy the material saves from mechanical work, from the passive starting point up to the elastic limit. In the same diagram, the area between points A, B, D, and C indicates the formability of the wire, and the total area under the slope from the starting point up to the failure point shows its modulus of toughness. Toughness is the total energy that can be stored from the mechanical work by the material, from the passive starting point up to the failure point.[3]

Wire performance

There are three properties that define the performance of a wire: stiffness, strength, and working range.

Stiffness (load/deflection rate)

Stiffness is the resistance of a wire to tension or bending. Wires with low stiffness have high elasticity, and their slope is close to horizontal. They can be bent easily, and they return to their original position when the stress is removed. Superelastic nickel titanium (NiTi) alloys are the best examples of these wires.

Wires with high stiffness show a steeper slope; high force is required to bend them. Stainless steel (SS) and

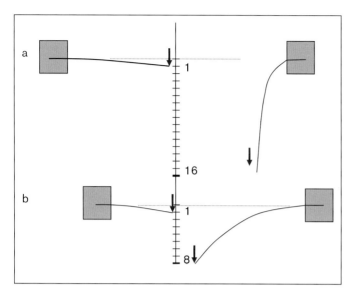

Fig 2-4 *(a)* The wire with half the size of the other shows 16 times more elasticity under the same force. *(b)* The longer wire (twice the length of the other) deflects 8 times more under the same force.

heat-treated chrome-cobalt alloys such as Elgiloy (Rocky Mountain Orthodontics) are examples of stiff wires. Generally, wires of low stiffness (high elasticity) are preferred at the first stage of treatment, and wires of high stiffness (low elasticity) are used at the final stage.

Stiffness or load/deflection rates of wires are affected by three factors: size, length, and material. To change the stiffness of a wire, one, two, or all three of these factors must be addressed.

Size In round wires, the force applied by the wire is directly proportional to the fourth power of the change in wire size. For example, when the wire size is doubled, the force applied increases 16 times. If the wire size is reduced to half its original size, the force is reduced 16 times.[7]

If forces of the same magnitude are applied to two wires of which one is twice the size of the other, the thinner wire deflects 16 times more than the thicker one (Fig 2-4a). This illustrates the impact that size has on the stiffness of wire. In rectangular wires, however, the force applied by the wire is directly proportional to the width (ie, the "edgewise" dimension) of the wire in making first-order bends and to the cube of its thickness (ie, the vertical dimension) in making second-order bends. This means that a wire of twice the width will deliver twice the force. A wire of twice the thickness will deliver eight times the force.

Length A force applied by a wire is indirectly proportional to the cube of the wire's length. This means that if the length of the wire is doubled, the force is reduced one-eighth; if the length of the wire is halved, the delivered force will be eight times as much. If equal forces are applied on wires of which one is twice the length of the other, the longer wire will deflect eight times that of the shorter wire (Fig 2-4b).[7] The purpose of the loops is to increase the wire distance between two brackets, thus increasing the elasticity to apply long-range, physiologic forces.

The interbracket distance is an important factor affecting the elasticity of the wire. Because the distance between narrow brackets is larger than the distance between wide brackets, the elasticity of the wire is higher where narrow brackets are used (Fig 2-5). This directly affects the amount of force delivered to the bracket by the wire. In narrow (single) brackets, the play between the bracket slot and the wire is larger than in wider (twin) brackets, and thus the force delivered by the wire is less (Fig 2-6 and Table 2-1).

This is particularly important during the initial phase of treatment, when the differences between bracket levels are considerable. At the start of the leveling phase, it is important to apply gentle forces to get cellular reaction in the periodontal tissues with tipping. Large interbracket distances and large second-order contact angles (the angles between the wire and the slot) make it possible to get faster leveling and alignment. To extrude a high canine, for example, the wire can be ligated at a single point to increase elasticity and working range and to avoid excessive moments on adjacent teeth (Fig 2-7).

Material The third factor that defines stiffness in a wire is the material from which it is made. Metal alloys have been used for several years in orthodontics. The most common wire alloys currently used are SS, cobalt-chrome-nickel (eg, Elgiloy), NiTi, and β-titanium (titanium-molybdenum alloy [TMA]). The stiffness of SS and heat-treated chrome-cobalt is very similar. The

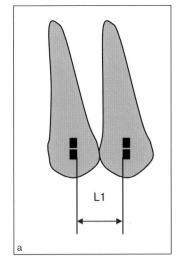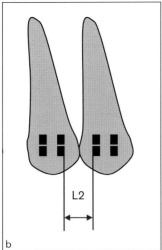 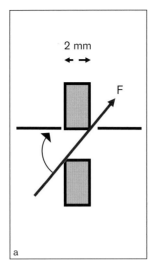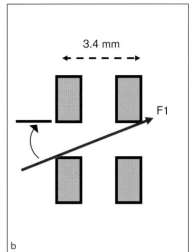

Fig 2-5 Interbracket distance is an important factor affecting the stiffness of the wire. The elasticity of the wire is higher with single brackets (a) because the distance between them is greater than the distance between the wide twin brackets (b). Therefore, the force applied to narrow brackets is less than the force applied to wide brackets.

Fig 2-6 The play of the wire in a narrow single bracket slot (a) is more than that in the wide twin bracket (b). The same size of wire would deliver less force (F) to the single bracket than the twin bracket (F1).

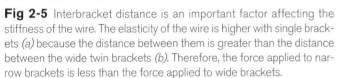

Wire size (inch)	Bracket width	0.018-inch slot (degrees)	0.022-inch slot (degrees)
0.016	Single narrow	1.15	3.43
	Medium twin	0.44	1.32
	Wide twin	0.32	0.95
0.017	Single narrow	0.57	2.86
	Medium twin	0.22	1.10
	Wide twin	0.16	0.80
0.018	Single narrow	0	2.29
	Medium twin	0	0.88
	Wide twin	0	0.64
0.019	Single narrow	–	1.72
	Medium twin	–	0.66
	Wide twin	–	0.48

*Reprinted from Kapila et al[8] with permission.
†In all combinations, the contact angles of the wire with the narrow brackets are larger than those with the wide brackets.

stiffness number of SS is always accepted as 1. NiTi and β-Ti wires have low stiffness compared with SS wires. Table 2-2 lists wires of different sizes and materials with equal bending stiffness.[9]

Strength

Strength, the second property that defines the performance of a wire, is the maximum force that the wire material can sustain. In a stress-strain diagram, the highest

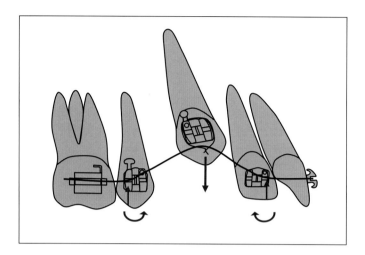

Fig 2-7 Interbracket distance has a substantial effect on wire stiffness. In leveling, if the difference between brackets is significant, the wire can be attached to the bracket at a single point to increase the length, thus increasing the elasticity and working range as well.

Table 2-2 NiTi, TMA, and SS wires of equal bending stiffness, according to their sizes*

	Bending		
NiTi	β-Ti (TMA)	SS	Relative springiness[†]
0.016	–	0.0175 (3 × .008)	6.6
0.019	0.016	0.012	3.3
–	0.018	0.014	1.9
0.017 × 0.025[‡]	–	0.016[†]	1.0
0.021 × 0.025	–	0.018	0.70
–	0.019 × 0.025	–	0.37
–	–	0.019 × 0.026	0.12

*Reprinted from Kusy[9] with permission.
[†]Scale of relative springiness, with 0.016-inch SS and 0.017 × 0.025–inch NiTi used as standard. The higher the numbers, the lower the springiness.
[‡]The 0.017 × 0.025–inch (rectangular) NiTi and 0.016-inch (round) SS wires have the same stiffness. This shows, clinically, that 0.017 × 0.025–inch NiTi can be used instead of 0.016-inch SS.[1]

Table 2-3 Comparison of strength, stiffness, and working ranges of 0.016- and 0.018-inch SS, TMA, and NiTi wires*[†]

	Strength	Stiffness	Range
SS	1.0	1.0	1.0
TMA	0.6	0.3	1.8
NiTi	0.6	0.2	3.9

*Reprinted from Kusy[9] with permission.
[†]The stiffness of TMA is one-third that of SS, while the stiffness of NiTi is only one-fifth that of SS. On the other hand, the strength of NiTi and TMA is approximately half that of SS wires.[1]

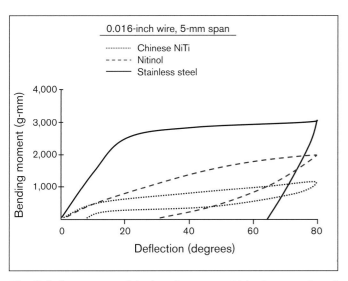

Fig 2-8 Comparison of the bending moment/elastic properties of 0.016-inch SS, austenitic Chinese NiTi, and Nitinol (martensitic NiTi) wires. Note that Chinese NiTi is twice as elastic as the Nitinol wire, and the springback property is also much higher. On the other hand, SS bent up the same amount (80 degrees) as the titanium alloy but can spring back only up to 64 degrees. (Reprinted from Burstone et al[10] with permission.)

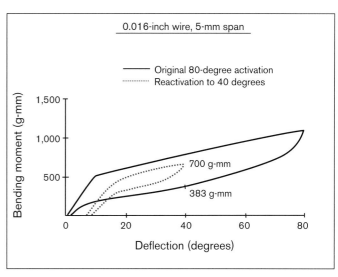

Fig 2-9 Result of activation of 0.016-inch austenitic Chinese NiTi wires, first up to 80 degrees and then to 40 degrees. The moment applied by the wire during deactivation is 383 g-mm. When the same wire is reactivated (removed from the bracket and then replaced), the force it produces at 40 degrees of activation is nearly doubled. (Reprinted from Burstone et al[10] with permission.)

magnitude of force applied to the wire on the *y*-axis shows the strength of that wire. Strength also defines the force-storing capacity of a wire.

Working range

Working range is the maximum elasticity of a wire material before permanent deformation occurs. On the stress-strain diagram, the distance between the projection of the elastic limit and the springback point after 0.1% permanent deformation of the wire on the *x*-axis shows the working range of that wire. The wires with high working range are those that can work for long periods with a single activation. Superelastic NiTi and TMA wires are good examples of wires with a high working range. Stiff wires such as SS and Elgiloy, however, have a relatively low working range. Table 2-3 compares the working range, stiffness, and bending strength of 0.016- and 0.018-inch SS, TMA, and NiTi wires.[1]

Apart from stiffness, strength, and working range, two more properties important in orthodontic practice are springback and formability.

Springback

When a wire is bent and set free within its elastic deformation limits, it returns to its original point. However, if that limit is exceeded, the wire cannot return to the original point. For the same amount of activation, the possibility of permanent deformation of SS and chrome-cobalt wires is higher than that of NiTi and TMA wires.

Springback is one of the most important criteria for defining the clinical performance of wires. The force delivered by the wire to the point of deactivation is critical from a clinical point of view. Figure 2-8 compares the elasticity and springback curves of SS, Nitinol (3M Unitek), and Chinese NiTi wires. For the same amount of activation (80 degrees), SS has the lowest springback value, whereas Chinese NiTi has the highest. The characteristic that makes Chinese NiTi wire unique is that after it is activated to be engaged in the bracket, then relaxed and reactivated, it delivers nearly twice as much force as before[10] (Fig 2-9).

2 | Application of Orthodontic Force

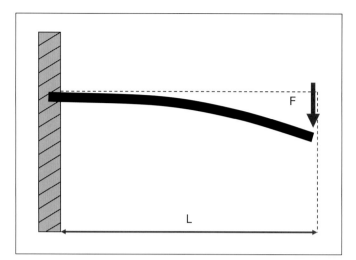

Fig 2-10 A cantilever beam consists of a wire of a certain length attached firmly to a block at one end. L, interbracket distance; F, force.

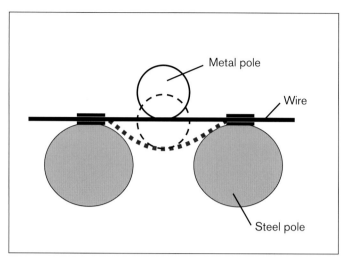

Fig 2-11 The three-point bending test recommended by Miura et al.[11] (Reprinted from Miura et al[11] with permission.)

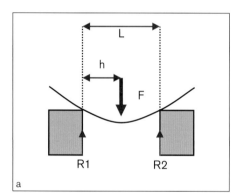
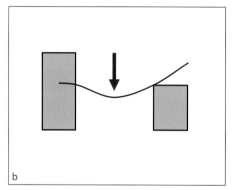
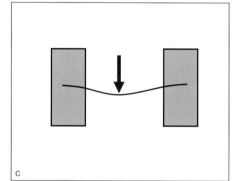

Fig 2-12 The beam type, used in the three-point bending test, wherein both ends of the wire are free (*a*); one end is free (*b*); and both ends are attached (*c*). Clinically, all types of beams are applicable, but the one with both ends attached seems to simulate most situations. L, interbracket distance; h, distance from the midpoint to the bracket; F, force. (Reprinted from Kusy and Greenberg[13] with permission.)

Formability

Formability is the area between the failure point of the wire and the permanent deformation limit on the stress-strain diagram (see Fig 2-2). This characteristic shows the amount of permanent deformation a material can sustain before it breaks.[1]

Testing the physical properties of materials

There are various testing methods used for determining the physical properties of wires. The simplest one is the cantilever beam test, which consists of applying force on a piece of wire of known size and length, attached firmly at one end to a block. In this test, when a force is applied to the wire from a certain point, a bending moment of $F \times L$ is generated (Fig 2-10). The wire resists bending in proportion to its stiffness, which is inversely proportional to the bending moment. This type of beam is an example of a cantilever-type, uprighting spring, but it cannot represent a multibracket system. The three-point bending test (Fig 2-11) is a better simulation of the wire-bracket relationship in a multibracket appliance.[11]

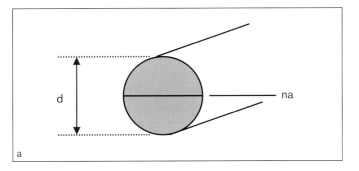
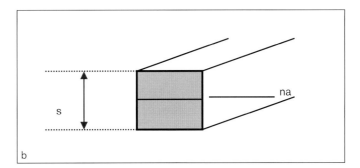
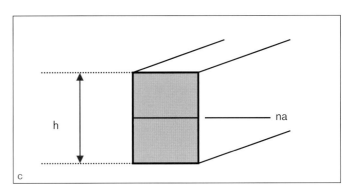

Fig 2-13 Formulas for the moments of inertia related to the cross-section of wires. (a) $I = \delta d^4/64$. (b) $I = s^4/12$. (c) $I = bh^3/12$. (Reprinted from Kusy and Greenberg[13] with permission.)

In this test, three combinations are possible from a clinical point of view:

- The state of free movement of the wire within two brackets (Fig 2-12a)
- The state of free movement of the wire in one bracket and attachment to the other bracket (Fig 2-12b)
- The state of the wire attached to a bracket at each end (Fig 2-12c)

The formulas vary according to the type of beam, but this general formula[12] is applicable to all three types:

$$\delta = L^3 \times F/N \times E \times I$$

where δ is the amount of deflection, L the interbracket distance, F the force causing the beam to deflect, N the stiffness of the beam, E the modulus of elasticity of the beam material, and I the moment of inertia of the beam material.

The value of N depends on the types of supports. In beams where both ends are free, N is 48 (see Fig 2-12a). In beams having both ends attached, this value is 192 (see Fig 2-12c).

The moment of inertia (I), as illustrated in Fig 2-13, is a physical parameter dependent on both the shape and size of the beam.[13,14] Therefore, it is easy to determine the moment of inertia of a beam of a certain shape and size. In multistrand wires, the moment of inertia of each strand must be calculated separately.[15] This is an important, variable factor when defining the elasticity of the beam. Increasing this value decreases the elasticity of the beam. In Fig 2-13, the round wire has the lowest moment of inertia, whereas the rectangular wire has the highest. Therefore, for the same beam type (the beam with both ends free, for example), the elasticity of the round wire will be higher compared with that of the rectangular wire.

Each type of beam has different values of stiffness, strength, and working range. These properties also depend on the length of the beam; thus, the interattachment distances affect the functions of all the beam types the same way. According to the formulas previously mentioned, the elasticity of the wire between brackets is directly proportional to the cube of the length of the wire between the two attachments (see the earlier section on factors affecting the stiffness of wires). The working range of the wire is directly proportional to the square of its length but inversely proportional to its strength.

The short and long edges of a rectangular wire have different moments of inertia. The moment of inertia of the first order is higher than that of the second order (ie, the bending stiffness of the first order is higher than that of the second order). Clinically, the problems in the first order (such as rotation, crossbite, scissors bite) must be corrected with flexible round wires before rectangular wires are inserted. The transverse dimension of the dental arch can then be maintained with rectangular wire.

Fatigue

Fatigue is the weakening (reduction of elasticity) of materials under repeated stress. The weakening effect of fatigue starts where the stress is concentrated. These are the points where the size of the object shows immediate changes such as fissures, crushes, notches, cracks, and welds of two materials. The points where the wire is crushed with pliers during bending and the areas of sharp bends are the most common points where fatigue failure occurs. Clinically, failure of wires with long interbracket distances (such as 2 × 4 archwires) is a common complication. No matter what the stiffness of the wire, the reason for failure is repeated occlusal contacts or chewing cycles. Failure usually occurs at sharp bends or the points of contact between the wire and the edges of the bracket slot. The way to avoid this is to bend the wire with round-tipped pliers or heat-treat the wire to homogenize its molecular structure. Chewing forces contribute to archwire failure, even within normal interbracket distance. This phenomenon can be explained by the deep notches at the points where NiTi wires contact the edges of the bracket slot.

Corrosion

From the time of manufacturing, the metallic materials used in orthodontic appliances are under the influence of physical and chemical agents related to their structural properties and environmental conditions. Corrosion is the change in mechanical properties and the metal's loss of weight under the effects of various chemical agents.

The oral environment, with its ions, carbohydrates, lipids, proteins, amino acids, and nonionic elements, is a suitable medium for the surface and deep abrasion of orthodontic attachments. Chlorine ions and the sulfuric compounds in the presence of microorganisms can corrode even SS appliances.[16,17] Food and beverages are effective in shifting the salivary pH toward acid or alkaline. Long-term accumulation of food around orthodontic appliances catalyzes the corrosion. Metals (restorations, wires, bands, and brackets) and molecular solids (elastics, cement, adhesive, and acrylics) are also affected by the oral environment.[3]

Matasa[16] and Maijer and Smith[17] have studied the corrosive effects of the oral environment on orthodontic appliances. They found that these effects can manifest themselves in any of the following corrosion defects: uniform, pitting, crevicular (crevice), intergranular, microbiologic, and electrochemical.

Uniform The metal surfaces of orthodontic appliances are uniformly exposed to corrosion. The loss of weight and mechanical characteristics are in proportion to the degree of exposure. Uniform corrosion is rarely seen on orthodontic attachments themselves, as these materials do not frequently come in contact with corrosive agents.[17]

Pitting Pitting corrosion is the most common type seen on orthodontic attachments. The mechanical characteristics and appearance of the materials are affected more than its weight. This type of corrosion is noticed mostly on materials that have welded or soldered joints and are not well polished. Parts that are not suitably manufactured or materials containing substances that compromise their purity are more prone to becoming corroded. Matasa[16] found that the chloride from salt ions is particularly responsible for this kind of corrosion.

Crevicular Crevicular corrosion is another common defect found on orthodontic attachments. It occurs especially in the presence of chlorides when the attachments are in contact with materials such as adhesives, acrylic, and elastics.[16,17] SS is considered to be especially sensitive to this kind of corrosion.[18]

Intergranular The appearance of the metal and its weight do not change during intergranular corrosion, but there is a loss of its mechanical characteristics, and failure may even occur. This insidious attack starts from inside and can reach the grains of the metal.[16] When SS is heated to a temperature of 400°C to 900°C, there will be a loss of chromium carbide at the granular borders, which renders the metal sensitive to this type of corrosion.[18]

Microbiologic The surfaces not in contact with air, such as the bracket base, may be affected by microbiologic corrosion.[19] Various microorganisms such as *Desulfovibrio desulfuricans* and *Desulfotomaculum*; oxidants such as *Thiobacillus ferroxidans* and *Beggiatoa;* and *Thiothrix, Aerobacter,* and *Flavobacterium;* produce sticky and humid matter, which affects the iron in SS. There are also iron-consuming microorganisms such as *Sphaerotilus, Hyphomicrobium,* and *Gallionella*.[16]

Electrochemical Saliva serves as a good medium for electrolytic reaction between metals. There is constant friction between the wire and the bracket slot, and this causes fretting corrosion on the metal surfaces in contact and possible appliance breakage. As a result of corrosion, heavy metals such as nickel, cobalt, and chromium deteriorate in the oral environment. This is particularly important for patients with sensitivity to nickel. To overcome the predisposition of metal alloys to corrode, titanium brackets and wires have been introduced.[16]

Force Elements Used in Orthodontic Appliances

To obtain quick and optimal tissue reaction, ideal force application is essential. For example, the force advised for successful canine distalization in sliding mechanics is approximately 150 to 200 g.[20,21] For this purpose, a variety of active elements such as coil springs, elastics, and loops can be used. An ideal force element in an orthodontic appliance should possess the following characteristics:

- Be able to deliver a constant and continuous force at optimum level
- Be hygienic and comfortable for the patient
- Be applied easily with minimum chair time
- Not be dependent on patient cooperation
- Be cost effective[21]

Wires

Stainless steel wires

SS wires are materials with high strength, high stiffness, low working range, and low springback properties.[1,13,22–24] High formability and low manufacturing costs have made them the most commonly used alloy for years. Because of their high stiffness, they are not suitable for the leveling stage. Their load/deflection rates must be reduced for them to be used in this period of treatment. To achieve this, either the length of the wire must be increased or its diameter reduced. However, reducing the load/deflection rate makes control of tooth movement more difficult.

The classic method of increasing wire length is to add loops. Because the force applied by the wire is inversely proportional to the cube of its length, the amount of force in looped wires decreases considerably. On the other hand, as their working range increases, the wires stay active for a long time.

Reducing the size of the wire is another method of decreasing the load/deflection rate, but this is limited to the sizes of wires on the market. Although reducing the size may increase the elasticity of the wire, it also increases the play between the bracket slot and the wire, which, in turn, results in loss of control over the tooth. For optimum force and movement control, the difference between the wire and the bracket slot size must be at least 0.002 inches.[1]

Multistrand wires Multistrand wires are manufactured by winding several thin SS wires onto one another. Their elasticity is relatively high, as their length increases with winding. The three- or five-strand round and eight- or nine-strand rectangular wires have been commonly used in clinical practice (Fig 2-14).

Loops

Loops have been used in fixed orthodontic treatment for many years. SS and chrome-cobalt are commonly used materials because they have a high range of bending stiffness. The main purpose of adding loops is to increase the elasticity of the wire by increasing its length. With the introduction of recently developed wires having high elasticity, strength, and wide working-range

 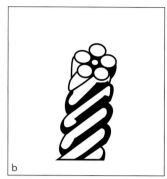 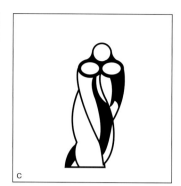

Fig 2-14 Types of multistrand wires.

properties (eg, NiTi and TMA), loops are no longer used as much.

Loops can have various shapes according to their purpose. They can be categorized as vertical, horizontal, combined, and special-shaped loops. It has been shown that the moment-to-force (M/F) ratio of a loop increases when more wire is added gingivally.[25,26] Loops are effective auxiliaries to open and close spaces and to correct inclinations and rotations if they are bent correctly and used properly. However, they can have certain disadvantages, too. First, because of increased elasticity, movement control might be more difficult. As soon as their "mission" is accomplished, they should be replaced with straight wire. Second, loop arches may cause hygiene problems and soft tissue irritation. Round wires should be especially avoided because they can easily roll within the bracket slot and irritate the gingiva. Irritation seldom occurs with rectangular wires. Even if it does happen, the wire can be bent outward with a finger while holding both sides of the loop with torque pliers to prevent bracket debonding.

Chrome-cobalt alloy wires

The physical characteristics of chrome-cobalt wires are very similar to those of SS.[13] Known on the market as Elgiloy, these wires consist of 40% cobalt, 20% chromium, 15% nickel, 7% molybdenum, and 15% to 20% iron. Elgiloy has different stiffness categories defined by four colors (ranked from the softest to the hardest): blue, yellow, green, and red. Soft (1.19) Elgiloy wire can be increased up to 1.22 with heat treatment.[24] Elgiloy wire is usually used in the Ricketts bioprogressive technique.

Nickel titanium wires

The properties of NiTi wires were discovered by Buehler in 1968, but their introduction to orthodontics and their development were realized later by Andreasen and Morrow.[23] The name of the first commercially available product, *Nitinol*, is derived from nickel titanium and Naval Ordnance Laboratory, where it was discovered during US space research.

Titanium alloy wires are marketed under the names *smart wires, shape memory wires, superelastic wires,* and others, but while some of them have these properties, most of them do not. It is necessary to understand the properties of these wires to make a proper selection and use them correctly.

Metallurgic properties NiTi wires are employed in either of two different crystal structures known as *martensite* and *austenite*,[27] depending on the temperature conditions in which they are used and the mechanical stresses acting on them. Austenite is seen in the higher temperatures and martensite in lower temperatures.[1,27] Austenite is a highly resistant, cagelike structure; martensite is actually a type of austenite that shows elastic shaping under heat and mechanical stress. The temperature at which austenite changes to martensite is called the *transition temperature*, and it is related mainly to the composition of the alloy. This process of conversion is called *martensitic transformation*, and it is reversible and can be repeated as much as needed.

Three main characteristics of NiTi wires make them different from SS and chrome-cobalt:

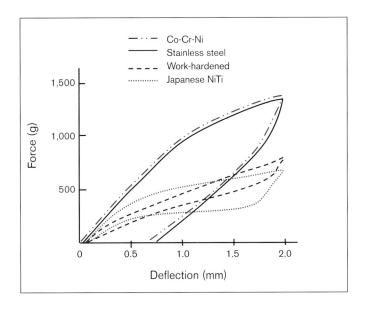

Fig 2-15 Comparison of superelastic Japanese NiTi (Sentalloy) wires with other wire alloys. Note that this wire applies more stable force compared with other wires. This is clinically important because constant and continuous force is desired to obtain optimal tooth movement. Co-Cr-Ni, cobalt-chrome-nickel. (Reprinted from Miura et al[11] with permission.)

- High elasticity
- Shape memory
- Resistance to permanent deformation[28]

NiTi wires are twice as elastic as SS wires, and their modulus of elasticity is 26% that of SS.[4] With these properties, NiTi alloys are ideal for leveling. However, as permanent deformation is directly related to time, some deformation can be expected while they are in the mouth.[4,10] Even though Nitinol has an extremely high springback property, because it is produced by a cool-hardening process, it does not have the superelasticity and shape memory of some of the other NiTi wires.[11] Shape memory is the shape-remembering process that allows the alloy to easily return to its original shape after being heated above a certain transition temperature.[29]

Superelasticity The term *superelasticity*, used for some NiTi wires, means that a wire can maintain a certain level of stress until a specific deformation point is reached and can keep that constancy on deactivation.[29] In other words, superelasticity is the force constancy that a wire can deliver, independent of the level of activation.[11,27] Miura et al[11] have shown that superelastic Japanese NiTi wire, marketed as Sentalloy (GAC), applies a fairly stable force (Fig 2-15). This force application has the desired physiologic properties for tooth movement and patient comfort.[29] Another superelastic wire, Chinese NiTi, shows 1.6 times more elasticity than Nitinol (see Fig 2-8).[10] During activation of a superelastic wire, when the stress reaches a certain level without a change of temperature, the wire converts from austenite to a martensitic structure and returns to austenite when the stress decreases to a certain level during deactivation.

A typical stress-strain diagram of superelastic wires is shown in Fig 2-16. When the wire is activated and the amount of force increases, the alloy transforms from the austenitic to the martensitic phase. When the force is removed, the wire follows a different path below that of activation, and returns to the austenitic structure. The difference between these curves is called *hysteresis*. This defines the difference between the force given for activation and the force applied by the wire during deactivation. This also shows the level of force a wire delivers to the tooth. Clinically, low hysteresis is desired. Light forces should be used, especially on incisors and premolars.[1,30,31] However, Segner and Ibe[32] showed experimentally that very few of the archwires tested deliver a clinically desired force level. Therefore, at the beginning of treatment, NiTi wires should be tied loosely to the brackets to avoid excessive force. This is important

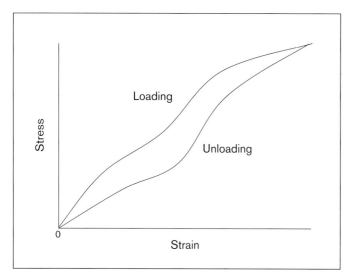

Fig 2-16 When a superelastic wire is activated, the material is transformed from the austenitic phase to the martensitic phase, and it follows a different path during deactivation. The difference between these two curves, called *hysteresis*, shows the clinical effectiveness of that wire.

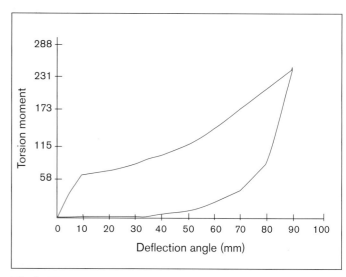

Fig 2-17 Hysteresis of thermally activated wires (Copper Ni-Ti) is low, and their springback is high. (Reprinted from Sachdeva[33] with permission.)

especially on the mandibular incisors with thin roots, where the interbracket distance is short.[32]

Superelastic wires remain in the austenitic structure at body temperature, changing to the martensitic phase when a mechanical stress is applied.[1,27] Copper Ni-Ti (Ormco) wires that become active at mouth temperature are also superelastic with shape memory. They are manufactured according to four transition oral temperatures (ie, 15°C, 27°C, 35°C, and 40°C). The difference between mouth temperature and the transition temperature determines the amount of force the wire will deliver. As this difference increases, the force delivered by the wire also increases. For example, in an oral environment of 36°C, the force delivered by a 15°C wire will be much higher than that of a wire of 40°C. These forces, that is, the transition temperature to be selected, will be determined according to the pain threshold of the patient and the desired duration of force (ie, continuous versus interrupted). For instance, the manufacturer recommends using 40°C wire for leveling an impacted canine. These wires, because of their high springback properties and low hysteresis, deliver very light forces, thus producing continuous tooth movement[33] (Fig 2-17).

Another characteristic making NiTi wires different is that they are not bendable, solderable, or weldable, like SS. Because of their high elasticity, loops are not practical on these wires. In daily practice, however, some small bends such as step-up, step-down, and stop bends can be made when necessary. Sharp or repetitive bends should be avoided to prevent breakage. Currently, some companies put midline stops on NiTi wires to prevent them from sliding through the slots and poking the cheek. If a cinchback is needed, it can be done by annealing the end of the wire with a clinical torch before placing it in the brackets.

Clinical performance From a clinical perspective, the effectiveness of superelastic wires can be a controversial topic. In a comparative study of patient discomfort over a 2-week period, no statistically significant difference was found between 0.014-inch superelastic Japanese NiTi wires and 0.0155-inch multistrand SS wires.[34] Jones et al[35] measured the velocity of crowding correction with a reflex metrograph between 0.014-inch Sentalloy and 0.0155-inch multistrand wires and found no statistically significant differences between the two materials. Similarly, in a comparative study between 0.016-inch superelastic Titanol (Forestadent) and conventional Nitinol, no significant difference in the velocity of tooth movement was found at the end of a 35-day period.[36]

β-Titanium molybdenum alloy wire

β-Ti (TMA) wires were introduced into the world of orthodontics in 1979 by Burstone and Goldberg.[37] Their elastic properties are between those of SS and NiTi. Despite their high elasticity, TMA wires are formable, solderable, and weldable—unlike NiTi alloys. Their modulus of elasticity is nearly twice that of Nitinol and one-third that of SS.[1,14,37] They have a broad working range as well as high biocompatibility.[38]

TMA wires have higher frictional values and surface roughness than SS and NiTi alloys. That is why they are commonly used in segmented arch mechanics rather than sliding mechanics. However, studies have shown that with ion implantation, the surface hardness of the material can be increased and surface roughness reduced to reach a frictional value comparable with SS wires. Because TMA wires with processed surfaces have twice the springback of SS, they can be used as straight leveling or finishing archwires.[39]

Wire selection in the clinic

In clinical orthodontics, selection of an archwire requires consideration of not only its physical properties but also factors such as the severity of malocclusion, desired type of tooth movement, and treatment methods.

Some of the properties expected from various wires used in orthodontic appliances are as follows:

- Must be sufficiently elastic; should not deform easily
- Must be formable and strong
- Should not be affected by oral fluids, acids, and other chemicals
- Should not corrode, rust, discolor, or oxidize
- Must be esthetic
- Should not be affected by heat or cold
- Must be inexpensive

Comparison of wire alloys

Kusy[40] and Kusy and Dilley[41] have studied the bending and torsional properties of SS, NiTi, β-Ti, and multistrand wires, with interesting results (Tables 2-4 to 2-6). Some of the results of these studies are as follows:

- The stiffness ratios of 0.012-inch SS to 0.018-inch NiTi and those of 0.014-inch SS to 0.018 × 0.018–inch NiTi are the same (0.8) (see Table 2-4). According to these values, from a stiffness point of view, a 0.018 × 0.018–inch NiTi can be used instead of a 0.012-inch SS wire. This result is important in two respects. The first is that the bending stiffness of the wire is related not only to its size but also to its material. The second is that rectangular NiTi wires, having a stiffness similar to round SS wires, can be used for the same purpose, such as leveling.

- The stiffness of rectangular wires was evaluated separately in the first and second order. According to Table 2-4, 0.016-inch SS is equivalent to 0.017 × 0.025–inch NiTi, and 0.018-inch SS is equivalent to 0.021 × 0.025–inch NiTi. Stiffness in the first order is higher than stiffness in the second order in all combinations (see the earlier section on moment of inertia). Second-order stiffness of 0.017 × 0.025 and 0.021 × 0.025–NiTi wires (the second order is possible only for rectangular wires) is 0.5 and 0.6, respectively. Because the stiffness of SS wires is accepted as 1, relative stiffness of NiTi wires is nearly half that of SS wires. Because their working range is 3.7 and 3.3 times that of SS wires, it is possible to achieve long-term tooth movement in the second order with NiTi wires.

- Table 2-5 shows the torsional properties of NiTi and TMA wires of the equivalent size to 0.019 × 0.026–inch SS wires. Although the strength ratios of NiTi and SS wires are very close to each other (ie, 0.8 and 0.9 for 0.019 × 0.025–inch NiTi and 0.021 × 0.025–inch NiTi, respectively), their working ranges are 5.4 and 5.3 times greater, respectively. The reason it is difficult to get torque with NiTi wires, even though they have strength values similar to SS wires, needs to be considered. Kusy[40] stated that NiTi wires could not easily deliver torque because of their low stiffness, high elasticity, and low formability.

In a study in which 0.0175-inch, multistrand (3 × 0.008–inch) SS wires and 0.018-inch solid SS wires were compared, few similarities, apart from their sizes, were found.[41] The solid SS has 11 times more stiffness and 4.6 times more strength than the multistrand wire (see Table 2-6).

The frictional values of multistrand wires are lower than those of many other wire alloys. The reason for this is the wavy surface topography and high flexibility of these wires, which prevents them from getting

Table 2-4	Comparison of elastic properties on bending of SS and NiTi wires*				
		Elastic property ratios		
SS	NiTi	Strength	Stiffness	Range
---	---	---	---	---
0.0175 inch (3 × 0.008)	0.016 inch	(1.7)	(0.9)	(1.9)
0.012 inch	0.018 inch	2.2	0.8	2.6
0.014 inch	0.018 × 0.018 inch	2.3	0.8	3.0
0.016 inch	0.017 × 0.025 inch	2.8[†]	1.1[†]	2.5[†]
		1.9[‡]	0.5[‡]	3.7[‡]
0.018 inch	0.021 × 0.025 inch	2.5[†]	0.9[†]	2.8[†]
		2.1[‡]	0.6[‡]	3.3[‡]

*Reprinted from Kusy[40] with permission.
[†]Edgewise.
[‡]Flatwise.

Table 2-5	Comparison of elastic properties on torsion of NiTi and TMA wires to 0.019 × 0.026-inch SS wires*			
	Elastic property ratios		
Wire type	Strength	Stiffness	Range
---	---	---	---
0.019 × 0.025-inch NiTi	0.8	0.1	5.4
0.021 × 0.025-inch NiTi	0.9	0.2	5.3
0.019 × 0.025-inch TMA	0.6	0.3	2.0

*Reprinted from Kusy[40] with permission.

jammed between the ligature and the bracket slot.[42] Multistrand wires are the wire of choice for leveling because of the reasons stated above and the fact that they are less expensive than NiTi wires. However, they are deformed easily by chewing forces, especially on long interbracket intervals.

Optimal tooth movement can be achieved easily with light and continuous force. Burstone[24] described four progressive levels of force during deactivation of the wire while moving an ectopic tooth into the arch. The force magnitude might initially be excessive, and then diminish as the tooth moves. It comes down to the optimal level, then to suboptimal and subthreshold levels, respectively (Fig 2-18). At the excessive force level, some indirect resorption of the surrounding bone may be observed. Activation of the wire decreases as the tooth moves, so the force comes down to the optimal level when optimal tooth movement is observed. With further movement of the tooth, activation is reduced even more, then tooth movement stops with complete deactivation. In practice, the orthodontist must try to keep the force at the optimal and suboptimal levels. This is why he or she must select the wire best suited for the desired tooth movement and magnitude of force.

Clinically, stiffness, or load/deflection rate, is one of the most important criteria in wire selection. The stiffness of a fixed appliance is basically defined by two factors. The first is related to the design of the appliance and the second to the wire. In other words, the stiffness of the appliance equals its inherent stiffness (depending on the appliance design) times the stiffness of the wire. The stiffness due to appliance design is directly related to factors such as the presence of loops or interbracket distance. Adding loops or increasing the interbracket distance (ie, using narrow brackets) reduces the stiffness of the appliance. The stiffness of the wire, however, is related to its configuration, size, length, and material.[24]

Table 2-6 Comparison of elastic properties of solid SS wires and 0.0175-inch, three-strand (3 × 0.008–inch) SS wires*

Solid SS	Bending Strength	Bending Stiffness	Bending Range
0.010 inch	0.78	1.0	0.76
0.012 inch	1.4	2.2	0.63
0.014 inch	2.2	4.0	0.54
0.016 inch	3.2	6.8	0.47
0.018 inch	4.6	11.0	0.42
0.020 inch	6.3	17.0	0.38
0.016 × 0.016 inch	5.5	12.0	0.47
0.017 × 0.017 inch	6.5	15.0	0.44
0.016 × 0.022 inch	10.0[†]	30.0[†]	0.34[†]
	7.5[‡]	16.0[‡]	0.47[‡]
0.018 × 0.022 inch	12.0[†]	34.0[†]	0.34[†]
	9.5[‡]	23.0[‡]	0.42[‡]
0.017 × 0.025 inch	14.0[†]	47.0[†]	0.30[†]
	9.6[‡]	22.0[‡]	0.44[‡]
0.018 × 0.025 inch	15.0[†]	50.0[†]	0.30[†]
	11.0[‡]	26.0[‡]	0.42[‡]
0.019 × 0.025 inch	17.0[†]	59.0[†]	0.29[†]
	13.0[‡]	31.0[‡]	0.40[‡]
0.021 × 0.025 inch	17.0[†]	58.0[†]	0.30[†]
	15.0[‡]	41.0[‡]	0.36[‡]

*Reprinted from Kusy and Dilley[41] with permission.
[†]Edgewise.
[‡]Flatwise.

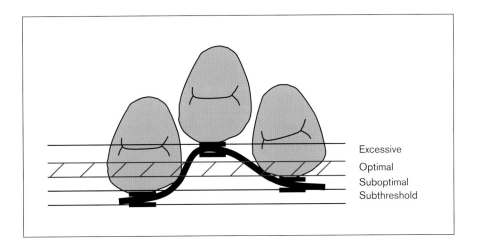

Fig 2-18 As an archwire ligated to a tooth out of the arch is deactivated, it applies four progressive force levels to the tooth. The excessive force magnitude when the wire is first attached will come down to the optimal and then to the suboptimal level as the tooth moves. When the tooth comes into the arch, the force would have fallen to the subthreshold level. (Reprinted from Burstone[24] with permission.)

Wire shape, size, and stiffness

Variable size orthodontics In orthodontic practice, the main criterion used to determine force levels has for many years been the size and shape of the SS wires. To maintain optimal force throughout treatment, orthodontists would use several wires of different shapes and sizes. This is called *variable size orthodontics*.[24] From time to time, the orthodontist modified the appliance design by adding loops to increase the elasticity of the wire. The force delivered by a wire is proportional to the fourth power of its size. This means that even a minor alteration in wire size can seriously affect the magnitude of the force delivered.

The complexity of wire-size formulas makes it difficult to determine suitable wire stiffness and wire selection. To illustrate the various options available, Tables 2-7 and 2-8 list the comparative stiffness of wires of different sizes and shapes.

According to these tables, the cross-section stiffness (CS) number of a 0.004-inch (0.102 mm) wire is accepted as 1, and all other wires are evaluated in relation to that. For example, the stiffness number of a wire with a 0.014-inch diameter is 150.06. This means that for the same amount of activation, a 0.014-inch wire will deliver 150.06 times more force than a 0.004-inch wire of the same material (see Table 2-7). Table 2-8 shows the CS numbers in the first and second order of rectangular wires. The orthodontist can confidently select the most suitable wire based on the information provided in these tables. It should be noted that the CS value increases as the wire size increases. This practice has been applied by orthodontists for years, and variable size orthodontics has subsequently evolved into a more practical and concrete methodology.

Variable modulus orthodontics The material properties of wires are defined with the modulus of elasticity (ie, the Young modulus). In wires of the same modulus of elasticity such as SS wires, size was the main criterion used to change the stiffness. However, with advancements in wire material technology providing different material properties and moduli of elasticity, it is now possible to have low wire stiffness and high working range in the same size wires, such as SS. This is called *variable modulus orthodontics*.[24]

Every wire has a CS and a material stiffness (MS). Multiplication of these two values gives the wire stiffness (WS) of that wire. The MS represents the stiffness value of the material. SS being the most commonly used material, its MS is designated as 1. Tables 2-9 and 2-10 show the WS numbers for solid wires of different sizes and materials and for multistrand wires, respectively.

For example, the stiffness of a 0.016-inch SS wire is 256, and the stiffness of an NiTi wire of the same size is 66.56, which is approximately one-fourth of 256. In practice, this signifies that a 0.016-inch NiTi wire is four times more elastic than an SS wire of the same size, or that for the same amount of activation, the NiTi applies one-fourth the force applied by the SS (see Table 2-9).

In variable modulus orthodontics, it is possible to change the stiffness values of the wires by changing the wire material without changing the size. For example, while the stiffness of a 0.016-inch SS and that of a 0.018 × 0.025–inch NiTi wire is nearly the same in the second order (256 and 251.38, respectively) (see Table 2-10), the stiffness of a 0.019 × 0.025–inch multistrand wire in the second order (78.46) (see Table 2-10) is less than that of a 0.012-inch solid SS wire (81.00) of the same size (see Table 2-9).

In practice, with two wires of the same stiffness, the larger wire is preferred to the smaller because it has more control over tooth movement. Because the heavier wires fill the bracket slot, there is less play in the slot compared with the thin wires. Therefore, the primary aim of the leveling phase is to start using thick wires as soon as possible to achieve more movement control. The same concept applies to loop arches.

The application of variable modulus orthodontics gives the clinician the opportunity to use wires of different materials with large working ranges that fill the bracket slot, providing movement control of the teeth right from the beginning. Therefore, the use of gradually increasing sizes such as 0.012 inch, 0.014 inch, 0.016 inch, and 0.018 inch is avoided. When the time comes to use conventional 0.016-inch SS wire, it will be possible to use 0.018 × 0.025–inch NiTi wire instead because the amount of force delivered by these two wires in the second order is the same.

Undoubtedly, a 0.018 × 0.025–inch wire has more movement control than a 0.016-inch round wire. Fur-

Table 2-7 Cross-section stiffness numbers of round wires*

Size (inches/mm)	CS
0.004 / 0.102	1.00
0.010 / 0.254	39.06
0.014 / 0.356	150.06
0.016 / 0.406	256.00
0.018 / 0.457	410.06
0.020 / 0.508	625.00
0.022 / 0.559	916.06
0.030 / 0.762	3.164.06
0.036 / 0.914	6.561.00

CS, cross-section stiffness.
*Reprinted from Burstone[24] with permission.

Table 2-8 Size stiffness numbers of rectangular and square wires*

Shape and size (inches/mm)	CS First order	CS Second order
Rectangular		
0.010 × 0.020 / 0.254 × 0.508	530.52	132.63
0.016 × 0.022 / 0.406 × 0.559	1129.79	597.57
0.018 × 0.025 / 0.457 × 0.635	1865.10	966.87
0.021 × 0.025 / 0.533 × 0.635	2175.95	1535.35
0.0215 × 0.028 / 0.546 × 0.711	3129.83	1845.37
Square		
0.016 × 0.016 / 0.406 × 0.406	434.60	
0.018 × 0.018 / 0.457 × 0.457	696.14	
0.021 × 0.021 / 0.533 × 0.533	1289.69	

CS, cross-section stiffness.
*Reprinted from Burstone[24] with permission.

thermore, the 0.018 × 0.025–inch NiTi wire has greater working range and strength than the 0.016-inch SS wire. This certainly has many advantages from a mechanical point of view, but the authors believe that when starting treatment our aim should be to stimulate tissue reaction around the teeth by applying gentle forces. Even though the rectangular wires mentioned above deliver low forces, they may generate uncontrolled and undesired moments right at the beginning of treatment and cause undesired side effects such as anchorage loss because of the so-called rowboat effect (see chapter 3).

This is especially true in cases where there is a serious level difference between the brackets and in crowding cases where there are inclination problems. In such cases, therefore, it is better to start treatment with round wires (such as 0.014-inch NiTi or 0.0155-inch twist-flex) that can easily roll in the bracket slot and move the teeth toward the arch with light tipping movements.

At the beginning of leveling, as the differences in level between the teeth are eliminated, the wire slides through the bracket slots and causes friction. Angulations between bracket and wire, in particular, may cause binding that

Table 2-9 Stiffness numbers of various solid wires with different cross-sections*

Wire type	Cross-section (inches/mm)	Order	MS†	CS	WS (MS × CS)
SS	0.009 / 0.229	NA	1.00	25.63	25.63
SS	0.012 / 0.305	NA	1.00	81.00	81.00
SS	0.014 / 0.356	NA	1.00	150.06	105.06
SS	0.016 / 0.406	NA	1.00	256.00	256.00
SS	0.018 / 0.457	NA	1.00	410.06	410.06
SS	0.020 / 0.508	NA	1.00	625.00	625.00
TMA	0.016 / 0.406	NA	0.42	256.00	107.52
NiTi	0.016 / 0.406	NA	0.26	256.00	66.56
TMA	0.018 / 0.457	NA	0.42	410.06	172.23
NiTi	0.018 / 0.457	NA	0.42	410.06	106.62
TMA	0.016 × 0.020 / 0.406 × 0.508	First	0.42	848.83	356.51
		Second	0.42	543.15	228.16
TMA	0.016 × 0.022 / 0.406 × 0.556	First	0.42	1129.79	474.51
		Second	0.42	597.57	250.98
SS	0.016 × 0.025 / 0.457 × 0.635	First	1.00	1865.10	1865.10
		Second	1.00	966.87	966.87
TMA	0.018 × 0.025 / 0.457 × 0.635	First	0.42	1865.10	783.34
		Second	0.42	966.87	406.08
NiTi	0.018 × 0.025 / 0.457 × 0.635	First	0.26	1865.10	484.93
		Second	0.26	966.87	251.38
EB	0.018 × 0.025 / 0.457 × 0.635	First	1.19	1865.10	2219.47
		Second	1.19	966.87	1150.57
EB-HT	0.018 × 0.025 / 0.457 × 0.635	First	1.22	1865.10	2275.42
		Second	1.22	966.87	1179.58
SS	0.021 × 0.025 / 0.533 × 0.635	First	1.00	2175.95	2175.95
SS	0.021 × 0.025 / 0.533 × 0.635	Second	1.00	1535.31	1535.31

NA, not applicable; EB, Elgiloy Blue; HT, heat treated.
*Reprinted from Burstone[24] with permission.
†The material stiffness of SS is arbitrarily set at 1.0 because it is the most commonly used wire material. Material stiffness (MS) × cross-sectional stiffness (CS) values give the wire stiffness (WS) number. For the same appliance, a given activation of 0.016-inch NiTi wire delivers approximately 0.26 times as much force as 0.016-inch SS.

tends to stop or delay tooth movement and cause anchorage loss. The play between 0.014- and 0.016-inch round wires and bracket slots is usually great enough to avoid binding. On the other hand, the amount of friction will certainly be higher if rectangular wires are used at this stage of treatment, especially NiTi and TMA wires that cause high friction because of their surface roughness. This is another important reason the use of rectangular wires should be deferred until the later stages of leveling. After obtaining acceptable leveling with round elastic wires, 0.016 × 0.022–inch or 0.019 × 0.025–inch superelastic NiTi wires can be applied for patient comfort.

Conclusion

SS and chrome-cobalt alloys have similar physical properties. They are still popular because of high strength, stiffness, formability properties, and low cost. They are

Table 2-10	Stiffness numbers of various multistrand wires with different materials and cross-sections*				
Wire type†	Cross-section (inches/mm)	Order	MS	CS	WS (MS × CS)
D-rect	0.016 × 0.022 / 0.406 × 0.559	First	0.036	1129.79	40.67
		Second	0.050	597.57	29.88
D-rect	0.018 × 0.025 / 0.457 × 0.635	First	0.048	1865.10	89.52
		Second	0.078	966.87	75.41
D-rect	0.019 × 0.025 / 0.483 × 0.635	First	0.056	1968.71	110.25
		Second	0.069	1137.13	78.46
D-rect	0.021 × 0.025 / 0.533 × 0.635	First	0.060	2175.35	130.56
		Second	0.065	1535.35	99.80
Respond	0.0175 / 0.4445	NA	0.069	366.36	25.28
Respond	0.0195 / 0.4953	NA	0.082	564.80	46.31
Respond	0.0215 / 0.5461	NA	0.068	834.69	56.76
Force-9	0.019 × 0.025 / 0.483 × 0.635	First	0.162	1968.71	318.93
		Second	0.135	1137.13	153.51
Hi-T Twist Flex	0.015 / 0.381	NA	0.175	197.75	34.61
Hi-T Twist Flex	0.0175 / 0.4445	NA	0.168	366.36	61.56
Hi-T Twist Flex	0.0195 / 0.4953	NA	0.153	564.80	86.41
Hi-T Twist Flex	0.0215 / 0.5461	NA	0.204	834.69	170.28

NA, not applicable.
*Reprinted from Burstone[24] with permission.
†All wires in this table are multistrand stainless steel. D-Rect (Ormco); Respond (Ormco); Force-9 (Ormco); Hi-T Twist Flex (3M Unitek).

excellent wires for some mechanics such as space closure, maintaining arch form, correction of axial inclinations, and torque control. Multistrand SS wires are preferred for leveling because of their high elasticity, wide working range, low frictional properties, and particularly low cost compared with NiTi wires. The disadvantage of those wires is that they are easily affected by chewing forces and can be deformed in moderate to severe crowding cases.

SS and chrome-cobalt alloys are not suitable for leveling. Putting a loop in a round wire is not a practical way to reduce stiffness because the wire can roll within the slot and irritate the gingival tissues. Straight, flexible wires work more effectively in leveling. However, loop bends still keep their importance, particularly in space closure.

NiTi wires, with their low stiffness and high elasticity, springback, strength, and working range, are excellent for leveling. These wires may also be used in the advanced phases of treatment for the correction of level differences between the teeth, such as those caused by accidental bracket failures. The fact that bending and soldering are not possible on NiTi wires limits their clinical use.

TMA wires, with stiffness properties ranging between those of SS and NiTi, are essential for frictionless, segmented arch mechanics owing to their high formability and wide working range.

2 | Application of Orthodontic Force

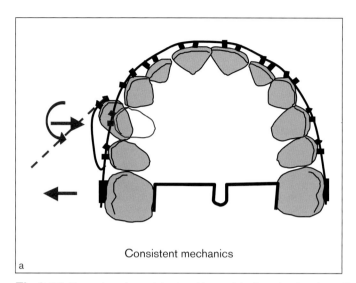
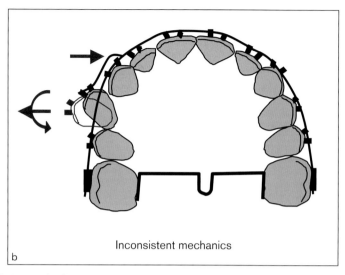

Fig 2-19 Examples of consistent and inconsistent mechanics. A cantilever between the first premolar and the archwire mesial to the molar tends to tip the molar buccally while rotating and tipping the premolar palatally into its place *(a)*. Changing the direction of the cantilever on the same premolar alters the mechanics. The premolar rotates but tips buccally as a reaction to palatally directed force in the canine area *(b)*.

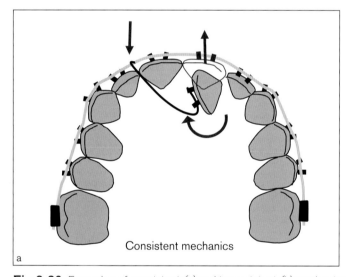
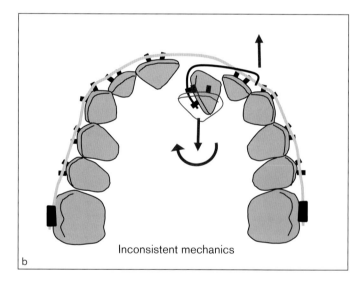

Fig 2-20 Examples of consistent *(a)* and inconsistent *(b)* mechanics.

Cantilevers

Cantilevers are useful tools for correcting occlusal plane inclination as well as individual tooth malpositions, not only in segmented arch mechanics but also as an auxiliary to straight-wire mechanics. Although several examples of cantilevers are given in this book, the mechanics need to be well understood to obtain optimum results.

The cantilever is nothing but a finger spring, such as that used in removable appliances, attached at one end to the acrylic (or engaged in the molar tube). The other end of the spring is free or attached at only one point. When activated, it applies a moment to the bracket on one end and a pure force to the other end. The mechanics of a cantilever are simple but effective. Two points need to be considered in applying cantilevers:

- The mechanics of the cantilever must be consistent with the direction of the desired tooth movement.
- The amount of force must be optimum to control anchorage.

Force Elements Used in Orthodontic Appliances

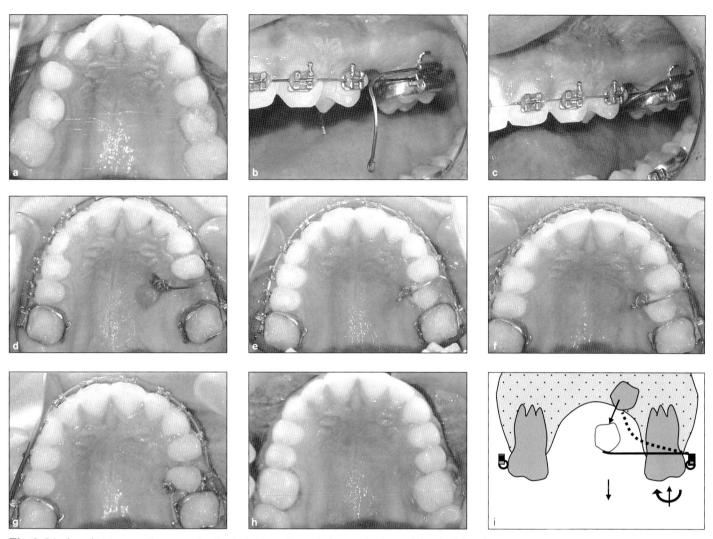

Fig 2-21 *(a to i)* Using cantilever mechanics to bring an impacted premolar toward the dental arch.

Figures 2-19 and 2-20 show examples of consistent and inconsistent mechanics. For cantilever mechanics to be consistent, the direction of the moment's balancing forces must move the teeth into proper positions on the dental arch. If necessary, anchorage should be reinforced to avoid adverse effects due to moment.

The advantage of cantilever mechanics is that they are predictable because the moments and forces are measurable. Cantilevers provide auxiliary mechanical aids to straight-wire mechanics for correcting severe rotations, moving ectopic teeth toward the arch, uprighting molars, and so on (Figs 2-21 and 2-22).

Coil springs

Because of their high elasticity, coil springs—especially those made from superelastic NiTi alloy—apply fairly constant and optimal force. Coil springs are made to increase the elasticity of wires by increasing their length. Open coils are activated by compression. They are used

39

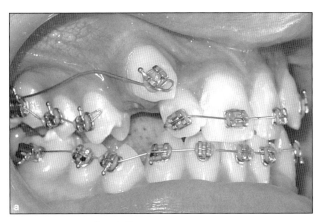 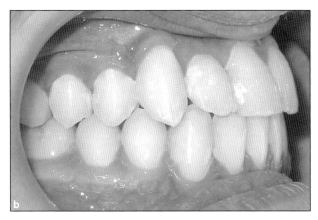

Figs 2-22a and 2-22b Cantilever mechanic used to bring a high canine to the arch. Note that the tip of the cantilever wire (0.016 × 0.022–inch) is connected to the canine bracket with only a single point.

to open spaces for correction of crowding, molar and premolar distalization/protraction, and space maintenance. Closed coil springs are activated by extension, and they are used mainly for space closure mechanics such as canine distalization and incisor retraction in sliding mechanics. Five factors affect the amount of force delivered by coil springs: alloy, lumen size, wire size, pitch angle of the coils, and length of the spring.[43]

Alloy

At present, coil springs are manufactured out of three alloys: SS, chrome-cobalt (Elgiloy), and NiTi. Studies to determine which material is the most suitable and which delivers a more constant force were done mostly in experimental environments simulating intraoral conditions.

Han and Quick[44] obtained the stress-strain graphs of NiTi and SS coil springs by stretching them up to three times their original length in synthetic saliva for periods of 2, 4, and 6 weeks. They found that the SS springs showed deformation after 2 weeks because of the environment and that they did not show any further deformation afterward. Conversely, the NiTi springs showed no alterations in their physical properties. When the SS and chrome-cobalt coil springs were compared, the latter were found to be stiffer than SS.[44]

In a comparative study of SS, chrome-cobalt, and Japanese NiTi springs, Miura et al[45] determined that the SS and chrome-cobalt closed coil springs showed a linear relationship and that the Japanese NiTi springs had a stable force value because of their superelastic properties. These wires did not show any permanent deformation, even when stretched up to five times their original size. In comparing open coil springs, they found that the forces applied by the SS and Elgiloy springs again showed a linear relationship, and that permanent deformation occurred when these wires were overcompressed. On the other hand, Japanese NiTi (again) had a constant force level without deformation.

Angolkar et al[46] compared the force degradation over time of SS, chrome-cobalt, and NiTi springs. They all showed force degradation, with the most significant loss of force taking place in the first 24 hours (Fig 2-23). The loss in the SS springs was 17.3% in the first 24 hours,

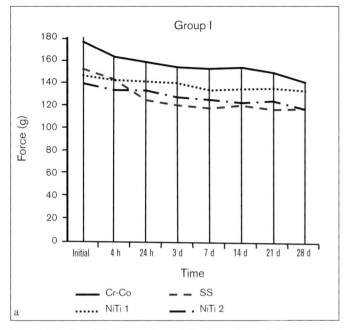
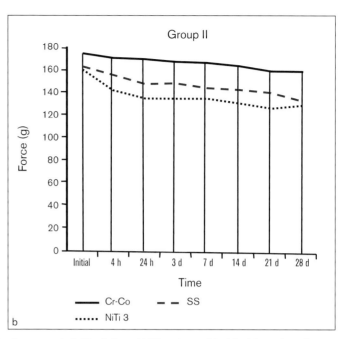

Fig 2-23 *(a and b)* Comparison of the force degradation over time of SS, chrome-cobalt (Cr-Co), and NiTi springs. (Modified from Angolkar et al[46] with permission.)

and that of the chrome-cobalt springs was 10%. The force degradation after 28 days for the three types of NiTi used in the study were: Ortho Organizers, 8.6%; Masel, 14.6%; and GAC, 17%. While the force degradation of the first two alloys is lower than the SS and Cr-Co springs, the force loss of the third NiTi alloy, contradicting the claims of Miura et al,[45] was higher than that of the other NiTi alloys and the same as that of the SS springs. This result, obtained by analysis of these superelastic NiTi alloys, cannot be fully explained by the researchers. We conclude that in comparing SS, Elgiloy, and NiTi springs, the NiTi have the lowest force degradation.

Lumen size
An increase in lumen size increases the length of the wire incorporated in it. This decreases the load/deflection rate.

Wire size
An increase in the size of the wire will add to the load/deflection rate, with a resulting decrease in elasticity.[46]

Pitch angle
Pitch angle is the angle between a perpendicular to the long axis of the spring and the inclinations of the windings. The winding number/length unit ratio decreases as this angle increases. Therefore, as the wire length decreases, the elasticity of the spring also decreases.[42,46]

Length of spring
An increase in the spring's length will decrease the load/deflection rate and increase its elasticity.[43]

Elastics

Elastic materials—some of the most frequently used active elements in orthodontic appliances—have been used increasingly for such movements as space closure in sliding mechanics, diastema closure, and rotation correction. Natural rubber and synthetic polymers are the two main elastic materials used in orthodontic practice. Natural rubber is a well-known material used for years in industry. In orthodontics, intramaxillary and intermaxillary elastics are usually made of natural latex.

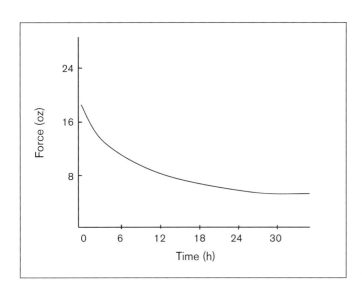

Fig 2-24 Force relaxation graph of an elastic material.

Synthetic polymers, now frequently used in orthodontic treatment as elastomeric chains, latex threads, and elastomeric modules, were developed from petrochemicals in the 1920s.[47] Synthetic elastics are amorphous polymers made of polyurethane.[48] Polymer chains slide over each other or they unfold when stretched by force. Sliding of the chains is a viscose movement; it is slow and irreversible. Unfolding of the chains by stretching is an elastic behavior, which is rapid and reversible. In orthodontic elastics, irreversible viscose behavior is observed toward the end of the application.[49]

In practice, the disadvantage of elastic materials is force degradation, which occurs over time. This process, called *relaxation,* is the loss or total elimination of the applied force when under constant stretching. Even if the amount of activation does not extend the elastic limits of the material, some loss of force due to relaxation is expected over time (Fig 2-24).[3]

Synthetic polymers are not ideal elastic materials because their mechanical properties are dependent on the usage time and the temperature.[48,49] They are only slightly affected when in contact with water for a short time, but after longer periods, they absorb water because of the hydrogen connections between their macromolecules and absorbed liquids,[49,50] which fill the spaces in the matrix. The discoloration of elastics in the oral environment is due to these liquids being taken into their structure.[47] Synthetic polymers are sensitive to ozone and ultraviolet rays, and their resistance and elasticity decrease under these influences. Manufacturers are trying to reduce this by adding antioxidants and antiozone elements and extend the shelf life of elastomers.[49]

Elastomeric chains

Studies of elastomeric chains show that at the end of the first 24 hours, these materials lose 50% to 70% of their original force. As a result of progressive force degradation, by the end of 3 weeks only 30% to 40% of the initial force remains.[50] Andreasen and Bishara[50] have stated that the greatest force degradation occurs during the first hour. Hershey and Reynolds[51] studied elastomeric chains of three different manufacturers and found no significant differences in force degradation, but significant differences were found in all their initial force levels. Researchers state that the correct clinical application is to determine the elastic's force with a dynamometer. The size of the elastomeric links (ie, short, medium, or wide) may also affect the chain's properties. The initial force application of wide elastomeric chains is light, but their force degradation is high.[47,48]

In a comparative study of gray and clear elastomeric chains, Williams and von Fraunhofer[52] have shown that the initial force of clear chains is higher and their force degradation is lower than gray elastomeric chains. It was also observed that fluoride-releasing elastomeric chains showed more force degradation than standard elastics.

Some researchers[47] recommend that elastomeric chains be stretched before application to prevent immediate force degradation and to apply force close to a constant level.

Hershey and Reynolds[51] state that the effect of heat increases the force degradation of elastomeric chains. In two studies on cold disinfection and sterilization of elastomeric chains, the materials were left in the solutions for periods of 30 minutes, 10 hours, and 1 week without affecting their characteristics.

Intraoral latex elastics

Latex is a natural material. In orthodontic practice, intramaxillary and intermaxillary elastics are made mostly of latex, marketed in various sizes and thicknesses according to the needs of a particular case. The most commonly used elastic sizes are 1/8-inch, 3/16-inch, 1/4-inch, 5/16-inch, and 3/4-inch. The amount of force listed on product packages indicates the amount of force in these elastics when they are stretched up to three times their lumen size. The force delivered by elastics of the same lumen size can vary depending on the manufacturer. For example, the 3/16-inch medium elastics produced by American Orthodontics deliver 4.5 oz, while Ormco's deliver 3.5 oz.

Andreasen and Bishara[50] have stated that the highest force degradation of latex elastics (approximately 40%) occurs on the first day. Bales et al,[53] analyzing whether the amount of force stated by the manufacturers can be obtained from the latex elastics when they are stretched up to three times their lumen sizes, observed that the elastics deliver higher force than expected. According to the study, a more suitable force level is obtained when they are stretched up to twice their lumen sizes and that latex elastics do not show significant differences in wet or dry environments.[53] The effects of the amount and duration of stretching in both wet and dry environments on force levels were tested for 1/4-inch medium elastics from four companies: GAC, Ormco, Dentaurum, and American Orthodontics.[54] Elastics were stretched up to three, four, and five times their lumen sizes and were placed in artificial saliva at 37°C for periods of 0, 1, and 24 hours, and 7 days. A group of identical elastics was placed in a dry environment at room temperature for the same periods as the first group. At the end of the predetermined time, they were tested with a universal testing machine, with the following findings:

- Environmental conditions have a significant effect on elastics' force degradation. The elastics placed in artificial saliva at 37°C showed more force degradation than the ones placed in a dry environment. The force degradation differences of the Ormco and American Orthodontics elastics placed in a dry environment (Fig 2-25) for 1 hour, 24 hours, and 7 days were statistically significant. In the wet environment (Fig 2-26), the force degradation differences between the Ormco, Dentaurum, and American Orthodontics elastics from the 0, 1-hour, 24-hour, and 7-day periods and the differences of the GAC elastics from 0, 24-hour, and 7-day periods were found to be significant.
- Force delivered by elastics increases as the stretching increases. However, the amount of stretching had no effect on the force relaxation of the elastics. For example, the force degradation of an elastic stretched up to three times its lumen size can be more than that of an elastic stretched up to five times its lumen size.[54]
- Length of time is an important factor in force degradation. In a wet environment, the force degradation differences of all the elastics at the beginning, 24 hours, and 7 days were found to be statistically significant.
- On the basis of percentage values when evaluated under equal conditions, the force degradation of the elastics of the two companies with low initial force was less than those of the two companies having high initial force.

Types of intraoral elastics Intraoral elastics are used basically as intramaxillary and intermaxillary. Intramaxillary elastics are applied as elastomeric chain or elastic thread between two attachments on the same arch. Chain elastics are the most popular force elements in canine distalization and incisor retraction in sliding mechanics. As the elastic force is applied to the buccal of the molars (not passing through the center of resistance), it may cause mesiopalatal rotation and some expansion (Fig 2-27).

Intermaxillary elastics are categorized according to the purposes for which they are used, such as Class II, Class III, vertical, triangular, anterior box, and crossbite. They are also categorized by their lumen sizes, such as 1/8-inch, 3/16-inch, 1/4-inch, or 5/16-inch, and by the amount of force they deliver, such as light, medium, heavy, and superheavy.

2 | Application of Orthodontic Force

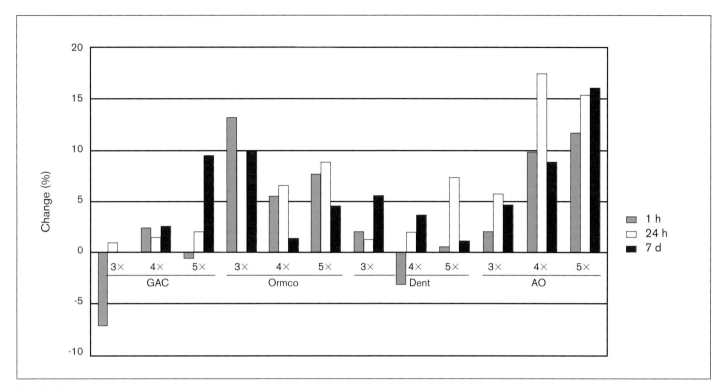

Fig 2-25 Changes as percentages of the elastics stretched up to three, four, and five times their lumen sizes and placed in a dry environment. Dent, Dentaurum; AO, American Orthodontics. (Reprinted from Dinçer et al[54] with permission.)

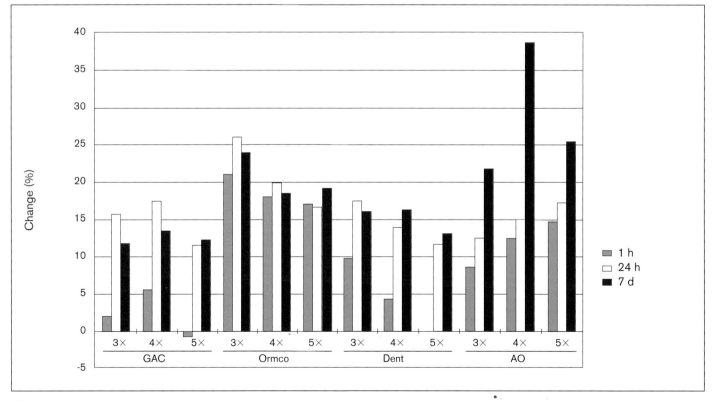

Fig 2-26 Changes as percentages of the elastics stretched up to three, four, and five times their lumen sizes and placed in a wet environment. Dent, Dentaurum; AO, American Orthodontics. (Reprinted from Dinçer et al[54] with permission.)

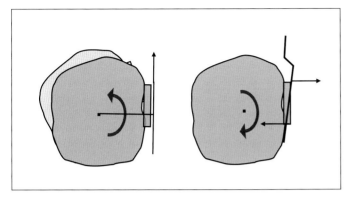

Fig 2-27 The molars rotate mesiopalatally from the effect of the force applied from the buccal. A toe-in bend in the wire will prevent this adverse effect.

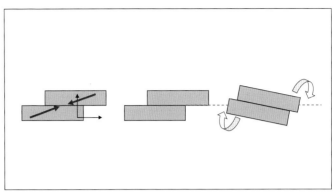

Fig 2-28 The mandibular molars and maxillary anterior teeth extrude from the effect of the vertical components of the Class II elastic force, causing clockwise rotation of the occlusal plane. This results in a deep bite and clockwise rotation of the mandible. (Reprinted from Langlade[56] with permission.)

The best method of selecting the optimal elastic is to measure, with a dynamometer, the force that the elastic delivers between attachments. As explained above, some force degradation in time is expected from relaxation of the material. Sometimes this takes place in the short time the patient opens and closes his or her mouth a couple of times. This is why some companies add an additional half-ounce of force to compensate for the initial force degradation and state in their catalogs that this value is soon lost.

Class II and Class III elastics The most commonly used elastics in daily practice are Class II and Class III. Class II elastics are applied between a mandibular molar hook and the maxillary canine or lateral incisor area. These elastics can be used for several purposes such as canine distalization, mandibular molar protraction, maxillary incisor retraction, and maxillary molar distalization. Some authors recommend distalizing canines on a round, 0.016-inch SS wire starting with a ¼-inch light elastic, continuing with a ¼-inch heavy elastic as the tooth moves on a 0.016 × 0.016–inch wire, and finish with a ³⁄₁₆-inch heavy elastic.[55] Molar protraction should be done on rectangular wires such as 0.016 × 0.022–inch or 0.017 × 0.025–inch SS, as these wires are stiff enough (assuming a 0.018-inch bracket slot) to resist mesial tipping of the molars.

Because Class II elastic force is applied diagonally between mandibular posterior and maxillary anterior teeth, it has both a vertical and a horizontal component (Fig 2-28). The vertical component tends to extrude the maxillary canines and incisors and the mandibular posterior teeth. Excessive use of heavy Class II elastics may cause the occlusal plane to rotate clockwise, moving the mandibular molars upward and the maxillary incisors downward, resulting clinically in a deep bite and clockwise rotation of the mandible.

Practically, the magnitude of the vertical component increases as the mouth opens. Figure 2-29 shows the increase in a Class II elastic's vertical component for a 25-mm mouth opening. The force values delivered by various Class II elastics on the maxillary and mandibular arches for different mouth openings are shown in Table 2-11.[56] In high-angle cases, molar extrusion due to excessive use of heavy elastics may cause clockwise rotation of the mandible, increase mandibular facial height, and worsen the soft tissue profile (Fig 2-30). To reduce this adverse effect, it is necessary to either reduce the vertical component or increase the horizontal component of the elastic force by engaging the elastic between the mandibular second molar and the hook either between the maxillary canine and lateral incisor (Fig 2-31) or on the lateral incisor. The point of application of the Class II elastic on the molar tube strongly

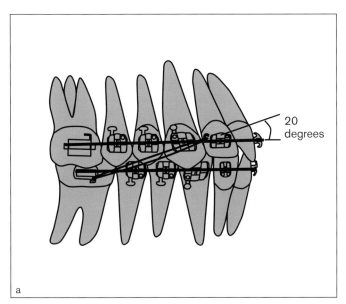 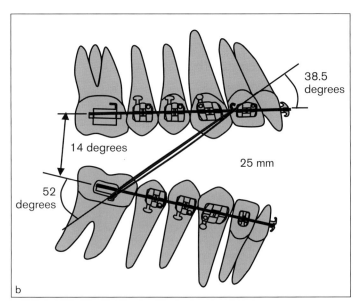

Fig 2-29 *(a)* In a closed mouth, the angle between the Class II elastics and the occlusal plane is approximately 20 degrees. The vertical component of force increases and the horizontal component decreases as the mouth is opened. *(b)* In a mouth opening of 25 mm, the angle between the elastic and the mandibular occlusal plane reaches 52 degrees. (Reprinted from Langlade[56] with permission.)

Table 2-11 Force values created by Class II elastics on the maxillary and mandibular dental arches when the mouth is closed and opened 10 mm and 25 mm*

Elastic	Opening (mm)	Distal force (g)	Mesial force (g)	Occlusal force (g)	Extrusive force (g)
1/4-inch light	in CO	78.0	78.0	28.4	28.4
	10	110.2	103.2	61.1	72.3
	25	124.4	96.8	99.0	126.1
1/4-inch heavy	in CO	94.0	94.0	32.2	34.2
	10	140.0	131.0	77.6	91.8
	25	148.7	115.7	118.3	150.7

CO, centric occlusion.
*Reprinted from Langlade[56] with permission.

affects molar movement (Fig 2-32). If the elastic is connected to the end of the archwire distal to the tube, the distance between the line of action of the force and the center of resistance increases, thus increasing the tipping moment. To avoid this, Class II elastics must be engaged on the molar tube's mesial hook.

Excessive wear of Class II elastics may result in extrusion, lingual tipping, and mesiolingual rotation of the molar (Fig 2-33). To prevent this, a toe-in bend can be placed in a rectangular SS wire and the elastic force reduced. The stiff rectangular wire will maintain the transverse dimension of the arch by eliminating the lingual tipping tendency of the molar.

Another undesirable adverse effect is protrusion of the mandibular incisors (Fig 2-34). Avoiding this is particularly important when mandibular incisor inclinations are critical and the frontal bone is thin. Further protrusion of the incisors may cause recession of the

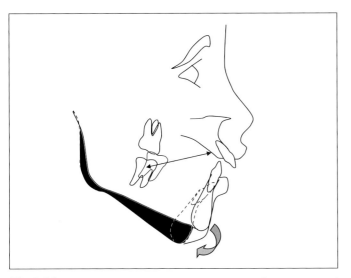

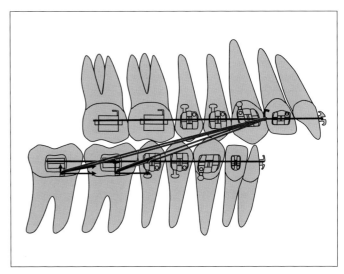

Fig 2-30 Excessive use of Class II elastics can result in extrusion of the mandibular molars, thus opening the bite. This is especially harmful in high-angle patients. As a result of molar extrusion, the mandible rotates clockwise, the chin moves downward and backward, and the profile becomes more convex.

Fig 2-31 To reduce the extrusive effect of Class II elastics, it is necessary to reduce the vertical component of force. The best way to achieve this is to place the elastic between the mandibular second molar and the distal aspect of the maxillary lateral incisor.

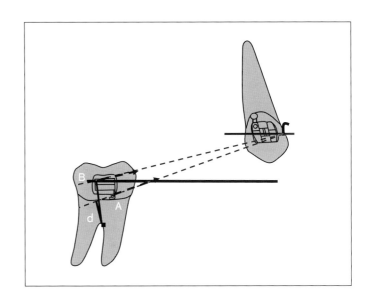

Fig 2-32 The point where the Class II elastic is engaged on the mandibular molar tube determines the type of tooth movement. To reduce tipping of the molar, the elastic must be engaged at the hook at the mesial of the tube (A). If the elastic is engaged at the back of the tube (B), the moment arm (d = distance) extends and the tipping effect increases.

labial gingiva. To prevent incisor protrusion, labial root torque can be placed in the rectangular SS wire[30] and wearing time and force level reduced.

In Class II or Class III patients in the transitional dentition, 2 × 4 mechanics can be used to obtain proper overjet and overbite. In the late transitional dentition, the use of Class II or Class III elastics after exfoliation of the second primary molar may cause the permanent first molars to tip forward and block the eruption path of the second premolars. If intermaxillary elastics are necessary, this space needs to be maintained. A step bend contacting the mesial aspect of the molar tube and the distal of the first premolar bracket can easily keep this space open until eruption of the second premolar (Fig 2-35a). Instead of step bends, an open coil spring can be placed between two brackets (Fig 2-35b).

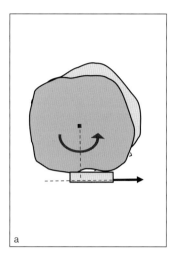 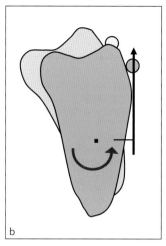

Fig 2-33 *(a)* The line of action of the elastic force applied to the mandibular molar passes buccally to the center of resistance. *(b)* This results in mesiolingual rotation and lingual tipping as well as extrusion.

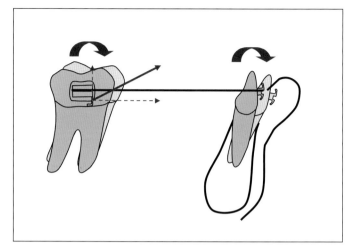

Fig 2-34 In the sagittal plane, the horizontal component of the Class II elastic force may cause protrusion of the mandibular incisors. The best way to avoid or minimize this (usually undesired) adverse effect is to use elastics for the minimum amount of time and place labial root torque in a rectangular SS wire.

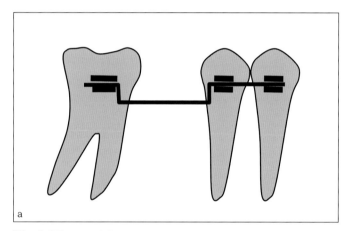 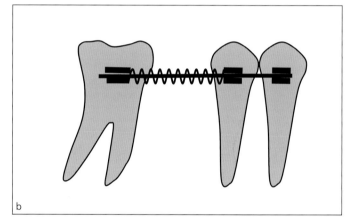

Fig 2-35 Use of Class II elastics in the late transitional dentition after the primary second molars have exfoliated may cause the permanent molars to tip forward and reduce the space available for the erupting mandibular second premolars. To maintain this space, step bends *(a)* or open coil wires *(b)* can be placed between the two teeth.

Springs versus elastics

Springs and elastics are the most commonly used active elements in space closure. In this section, the continuity of force and force degradation of these materials due to time, tension, and environment are compared.

The force characteristics of springs and elastics have been the subjects of several studies. Samuels et al[57] have analyzed the space-closing effectiveness of NiTi springs and elastomeric modules. These materials remained on the teeth for weeks without being changed. At the end of a 4-week period, researchers observed that the NiTi springs delivered a much more constant force that is better tolerated by the biologic tissues compared with the use of elastics. Sonis et al[58] clinically studied 150-g Sentalloy springs and 3/16-inch intraoral elastics that apply 180 g of force. They pointed out that in space closure, springs realize the tooth movements twice as fast as elastics, and they claim that this difference is due to the springs delivering more constant force than the elastics and that they do not require patient cooperation.

When they compared force degradation over time, they found that the SS materials showed the highest force degradation, with a 21.4% loss at the end of the 28-day period. The force degradation of NiTi wires from three different companies was 8.6% (Ortho Organizers), 14.6% (Masel), and 17% (GAC).[46] Meanwhile, the force degradation of the elastics after 1 month was only 50% to 60%.[51,59] According to these results, the elastics showed much higher force degradation than the springs. This loss of force makes it impossible to maintain the optimum amount of force throughout the whole movement. This supports the idea that elastics deliver interrupted force.

When kept in the mouth for a long time, elastic materials such as elastomeric chains and elastic ligatures become deformed, change color, and lose their force as they absorb liquids from the oral environment. In terms of oral hygiene, coil springs are not necessarily better than elastic materials because they trap food between the coils, whereas elastics, when changed regularly, do not have a negative impact on overall oral hygiene.

Elastic ligatures versus wire ligatures

Because of the flexibility of elastic ligatures, the bracket-wire relationship is relatively loose, which may be suitable at the beginning of leveling when one desires to stimulate tooth movement with light forces. As treatment progresses, however, especially in the final phase, it is usually preferred to use stiff rectangular wires securely held in the bracket slots to transfer all the force to the teeth. At that point, therefore, elastic ties should be replaced with SS wire ligatures. Despite the disadvantages of elastic materials compared with wire ties, elastics are presently preferred by most orthodontists because of their color options.

Attachments Used in Orthodontic Appliances: Brackets

Brackets are the most important elements of orthodontic appliances. They are classified according to their size: large, mini, and ultramini. Base structure may be straight or curved; width may be narrow, medium, or wide, or they may be single or twin, according to the technique used. The material may be SS, ceramic, plastic, composite, or titanium. The manufacturing techniques may be milled or machined, cast or sintered (metal injection molding). Instead of well-known features such as sizes or base structures, bracket materials and manufacturing techniques are described here.

Bracket materials

The structural characteristics of brackets are paramount for clinical efficiency. The bracket material must be hygienic, nontoxic, and resistant to corrosion. In addition, it must resist the forces applied to it by the wire or the occlusal forces, and it must be economical. Bracket materials should:

- Be esthetic
- Not absorb water
- Not be discolored by oral liquids
- Have minimal bracket-wire friction

SS brackets

Most brackets currently used are made of austenitic SS containing 18% chrome and 8% nickel. The standards are set by the American Iron and Steel Institute, with codes such as 303, 304, or 316, depending on the manufacturing companies. The most frequently used steel contents in bracket manufacturing are listed in Table 2-12.[15]

SS brackets have most of the basic characteristics expected from a bracket. Its resistance to all kinds of corrosion, hygienic properties, and reasonable price have made SS the world's most commonly used bracket material for many years. However, SS brackets have two important disadvantages: they are not esthetic, and they may release nickel into the oral environment. Various alternative materials have recently been developed to eliminate the esthetic problems, including ceramic, plastic, and composite materials.

It has been shown in various in vitro studies that SS releases nickel and chrome intraorally.[18,60,61] The amount of nickel released per day in the mouth of orthodontic patients is 40 µg, and the amount of chrome is 36 µg.[62] However, Bishara et al[63] found that even though nickel is released into the oral environment, the amount reaching the blood is very low. Nickel is known

Table 2-12 Types and composition of steels used in bracket manufacturing*

Designation			Composition (%)†							
AISI	UNS	DIN	C	Mn	Si	Cr	Ni	P	S	Other
303	S-30300	14305	0.15	2.0	1.0	17–19	8–10	0.20	0.02	0.6 Mo
304	S-30400	–	0.08	2.0	1.0	18–20	8–10.5	0.04	0.03	–
304 L	S-30403	14306	0.03	2.0	1.0	18–20	8–12	0.04	0.03	–
316	S-31600	14401	0.08	2.0	1.0	16–18	10–14	0.04	0.03	2–3 Mo
316 L	S-31603	14404	0.03	2.0	1.0	16–18	10–14	0.04	0.03	2–3 Mo
317	S-31700	14438	0.08	2.0	1.0	18–20	11–15	0.04	0.03	3–4 Mo
630/17-4 PH	S-17400	14542	0.07	1.0	1.0	15.5–17	3–5	0.04	0.03	4 Cu, 3 Nb
631/17-4 PH	S-17700	–	0.09	1.0	1.0	16–18	6.5–7.5	0.04	0.04	0.8–1.5 Al
ASTM-A 669	S-31803	–	0.0	1.0	0.5	22	5.5	0.02	0.02	3 Mo

*Reprinted from Matasa[16] with permission.
AISI, American Iron and Steel Institute; UNS, unified numbering system; DIN, Deutsches Institut für Normung.
†Balance is iron.

to cause various allergic reactions such as dermatitis and asthma,[61,63] so an alternative material must be used in sensitive patients. Manufacturers warn orthodontists and patients about this characteristic of SS material by including written notices on their packages.

Ceramic brackets

Ceramic is a biocompatible bracket material. It is also an ideal material because of its hardness and its esthetic, hygienic, and tissue-friendly characteristics. Ceramic brackets on the market have one of three structural characteristics: monocrystalline alumina, polycrystalline alumina, or zirconium. Alumina is harder than SS, but the failure stiffness of SS is 20 to 50 times more than that of ceramic.[64] Monocrystalline alumina brackets (eg, Radiance, American Orthodontics) are more resistant with smoother slot surfaces compared with other ceramic brackets.[65] On the other hand, polycrystalline alumina brackets (eg, Transcend, 3M Unitek) have rough surfaces, and their wings can be broken under uncontrolled torque stress.[65,66] Zirconium brackets are four times stronger than polycrystalline alumina.[64]

The fact that ceramic brackets are bulkier and more expensive than SS brackets and are breakable with high couples limits their use. In addition, ceramic material causes high friction during sliding mechanics. Therefore, some manufacturers produce brackets with SS inserted slots (eg, Clarity, 3M Unitek) to reduce friction between bracket and wire.

Akgündüz[67] studied the torque resistance of 280 polycrystalline (Fascination, Dentaurum; 20/40, American Orthodontics; Eclipse, Masel) and monocrystalline alumina (Starfire), standard, preadjusted edgewise (Roth) central and lateral brackets with either 0.018-inch or 0.022-inch slots against couples (palatal root torques) with full-size SS wires. Breakage was studied under a scanning electron microscope, and specific gravity was evaluated. Additionally, the resistance of monocrystalline alumina brackets to torque was analyzed with finite element analysis. The results are summarized as follows:

- The resistance of monocrystalline brackets was significantly higher than that of polycrystalline. It was noted that neither of the analyzed bracket types broke, even under 90-degree torque, while the SS wires were noticeably deformed.
- The breakage occurred on the gingival wings. This was supported by the finite element analysis, in which the highest stresses were observed at the corner of the slot where it meets the gingival wing.
- The stress on the brackets with rounded slot corners (Dentaurum) was less than that of the brackets from the other manufacturers.

- Even though the grains of the Dentaurum brackets were larger than those of the American Orthodontics and Masel brackets, their resistance was also higher because their specific gravity was higher than that of the other brackets.
- The resistance of the monocrystalline alumina (Starfire) brackets with high specific gravity was higher than that of the polycrystalline brackets having low specific gravity and porosities in their structure.
- The resistance of a bracket wing against torque was proportional to the thickness of the material between the slot and the section where the ligature passes. Brackets thicker in that area were more resistant.
- Analysis of resistance in the 0.018-inch and 0.022-inch brackets found them not significantly different in either the experimental study or the finite element analysis.

Plastic brackets

Plastic brackets are manufactured from hard polycarbonate material. This material has been reinforced with fibers to increase its hardness. These brackets are also esthetic and more economical than ceramic brackets, but they usually discolor in the oral environment and become dull yellow or gray as they absorb mouth liquids. Another disadvantage of plastic brackets is that they cause higher friction in sliding mechanics, although this problem has been alleviated by the manufacturer with the insertion of metal slots into the plastic body.

Conclusion

The fabrication of an orthodontic appliance must consider its parts and their material properties, composition, and behavior in the oral environment. For example, the frictional resistance in sliding mechanics may be less in sintered SS brackets than in cast or milled SS brackets,[68] and ceramic and polycarbonate brackets offer greater resistance than SS brackets.[69,70] Archwires are also available in a variety of materials, shapes, and sizes. Nitinol and temperature-sensitive wires with a long working range have increased the time interval required between appointments; multistrand SS wires are less expensive and have also increased the working range.[71] Finally, the structure and material properties of coil springs and elastics impact their rate of force degradation over time. Therefore, it is highly important to make careful decisions, based on the specific requirements of a case, regarding the components used to apply force in orthodontic treatment.

References

1. Proffit WR, Fields HW, Ackerman JL, Thomas PM, Tulloch JFC. Contemporary Orthodontics. St Louis: Mosby, 1986:238, 250.
2. Thurow RC. Technique and Treatment with the Edgewise Appliance. St Louis: Mosby, 1962.
3. Nikolai RJ. Bioengineering Analysis of Orthodontic Mechanics. Philadelphia: Lea & Febiger, 1985:104–109.
4. Burstone CJ. Application of bioengineering to clinical orthodontics. In: Graber TM, Swain BF (eds). Orthodontics: Current Principles and Techniques, ed 3. St Louis: Mosby, 1985:193–227.
5. Jarabak JR, Fizzell JA. Technique and Treatment with the Light-Wire Appliances. St Louis: Mosby, 1963.
6. Nanda R, Kuhlberg A. Principles of biomechanics. In: Nanda R (ed). Biomechanics in Clinical Orthodontics. Philadelphia: Saunders, 1996:12.
7. Philippe J, Aloé P, Sueur S. Orthodontie, des principes et une technique. Paris: Prélat, 1972.
8. Kapila S, Angolkar PV, Duncanson MG, Nanda RS. Evaluation of friction between edgewise stainless steel brackets and orthodontic wires of four alloys. Am J Orthod Dentofacial Orthop 1990;98:117–126.
9. Kusy RP. On the use of nomograms to determine the elastic property ratios of orthodontic archwires. Am J Orthod 1983;83:374–381.
10. Burstone CJ, Qin B, Morton JY. Chinese NITI wire. A new orthodontic alloy. Am J Orthod 1985;87:445–452.
11. Miura F, Mogi M, Ohura Y, Hamanaka H. The super-elastic property of the Japanese NiTi alloy wire for use in orthodontics. Am J Orthod Dentofacial Orthop 1986;90:1–10.
12. Adams DM, Powers JM, Asgar K. Effects of brackets and ties on stiffness of an arch wire. Am J Orthod Dentofacial Orthop 1987;91:131–136.
13. Kusy RP, Greenberg AR. Effects of composition and cross section on the elastic properties of orthodontic wires. Angle Orthod 1981;51:325–341.
14. Odegaard J, Meling T, Meling E, Holte K, Segner D. An evaluation of the formulas for bending with respect to their use in estimating force levels in orthodontic appliances. Kieferorthop Mitt 1995;9:73–88.
15. Kusy RP, Stevens LE. Triple-stranded SS wires. Evaluation of mechanical properties and comparison with titanium alloy alternatives. Angle Orthod 1987;57:18–32.
16. Matasa CG. Attachment corrosion and its testing. J Clin Orthod 1995;29:16–23.

17. Maijer R, Smith DC. Corrosion of orthodontic bracket bases. Am J Orthod 1982;81:43–48.
18. Park HY, Shearer TR. In vitro release of nickel and chromium from simulated orthodontic appliances. Am J Orthod 1983;84:156–159.
19. Matasa CG. Stainless steels and direct bonding brackets. III: Microbiologic properties. Inf Orthod Kieferorthop 1993;25:269.
20. Storey E, Smith R. Force in orthodontics and its relation to tooth movement. Aust Dent J 1952;56:11–18.
21. Reitan K. Some factors determining the evaluation of forces in orthodontics. Am J Orthod 1957;43:32–45.
22. Kusy RP, Dilley GJ, Whitley QJ. Mechanical properties of stainless steel orthodontic archwires. Clin Mater 1988;3:41–59.
23. Andreasen GF, Morrow RE. Laboratory and clinical analyses of nitinol wire. Am J Orthod 1978;73:142–151.
24. Burstone CJ. Variable-modulus orthodontics. Am J Orthod 1981;80:1–16.
25. Burstone CJ. The segmented arch approach to space closure. Am J Orthod 1982;82:361–378.
26. Manhartsberger C, Morton JY, Burstone CJ. Space closure in adult patients using the segmented arch technique. Angle Orthod 1989;59:205–210.
27. Drescher D, Bourauel C, Thier M. The materials engineering characteristics of orthodontic nickel-titanium wires [in German]. Fortschr Kieferorthop 1990;51:320–326.
28. Tosun Y. Biomechanical Principles of Fixed Orthodontic Appliances. İzmir, Turkey: Aegean University, 1999:39.
29. Viazis AD. Clinical applications of superelastic nickel titanium wires. J Clin Orthod 1991;25:370–374.
30. Ricketts RM. Bioprogressive Therapy. Denver: Rocky Mountain Orthodontics, 1979.
31. Rygh P. Orthodontic forces and tissue reactions. In: Thilander B, Rönning O (eds). Introduction to Orthodontics. Stockholm: Tandläkarförlaget, 1985:205–224.
32. Segner D, Ibe D. Properties of superelastic wires and their relevance to orthodontic treatment. Eur J Orthod 1995;17:395–402.
33. Sachdeva RCL, Miyazaki S. Biomechanical considerations in the selection of NiTi alloys in orthodontics and variable transformation temperature orthodontics with copper NiTi. In: Sachdeva R (ed). Orthodontics for the Next Millennium. Glendora, CA: Ormco, 1997:227–246.
34. Jones M, Chan C. The pain and discomfort experienced during orthodontic treatment: A randomized controlled clinical trial of two initial aligning arch wires. Am J Orthod Dentofacial Orthop 1992;102:373–381.
35. Jones ML, Staniford H, Chan C. Comparison of superelastic NiTi and multistranded stainless steel wire in initial alignment. J Clin Orthod 1990;24:611–613.
36. O'Brien K, Lewis D, Shaw W, Combe E. A clinical trial of aligning archwires. Eur J Orthod 1990;12:380–384.
37. Burstone CJ, Goldberg AJ. Beta titanium: A new orthodontic alloy. Am J Orthod 1980;77:121–132.
38. Williams DF, Roaf R. Implants in Surgery. Philadelphia: Saunders, 1973:315–318.
39. Burstone CJ, Farzin-Nia F. Production of low-friction and colored TMA by ion implantation. J Clin Orthod 1995;29:453–461.
40. Kusy RP. Comparison of nickel-titanium and beta titanium wire sizes to conventional orthodontic arch wire materials. Am J Orthod 1981;79:625–629.
41. Kusy RP, Dilley GJ. Elastic property ratios of a triple-stranded stainless steel arch wire. Am J Orthod 1984;86:177–188.
42. Tosun Y, Unal H. Study of friction between various wire and bracket materials. Turk Ortodonti Derg 1998;11:35–48.
43. Boshart BF, Currier GF, Nanda RS, Duncanson MG Jr. Load-deflection rate measurements of activated open and closed coil springs. Angle Orthod 1990;60:27–32.
44. Han S, Quick DC. Nickel titanium spring properties in a simulated oral environment. Angle Orthod 1993;63:67–72.
45. Miura F, Masakuni M, Ohura Y, Karibe M. The superelastic Japanese NiTi alloy wire for use in orthodontics. Part III: Studies on Japanese NiTi alloy coil springs. Am J Orthod Dentofacial Orthop 1988;94:89–96.
46. Angolkar PV, Arnold JV, Nanda RS, Duncanson MG Jr. Force degradation of closed coil springs: An in vitro evaluation. Am J Orthod Dentofacial Orthop 1992,102:127–133.
47. Baty DL, Storie DJ, von Fraunhofer JA. Synthetic elastomeric chains: A literature review. Am J Orthod Dentofacial Orthop 1994;105:536–542.
48. Young J, Sandrik J. The influence of preloading on stress relaxation of orthodontic elastic polymers. Angle Orthod 1979;49:104–109.
49. De Genova DC, McInnes-Ledoux P, Weinberg R, Shaye R. Force degradation of orthodontic elastomeric chains—A product comparison study. Am J Orthod 1985;87:377–384.
50. Andreasen GF, Bishara S. Comparison of alastik chains with elastics involved with intraarch molar to molar forces. Angle Orthod 1970;40:151–158.
51. Hershey G, Reynolds W. The plastic module as an orthodontic tooth-moving mechanism. Am J Orthod 1975;67:554–662.
52. Williams J, von Fraunhofer JA. Degradation of the Elastic Properties of Orthodontic Chains [thesis]. Louisville, Kentucky: University of Louisville, 1990.
53. Bales RT, Chaconas JS, Caputo AA. Force-extension characteristics of orthodontic elastics. Am J Orthod 1977;72:296–302.
54. Dinçer B, Erdinç AA, Tosun Y. Study of force relaxation in intraoral elastics. Turk J Orthod 2000;13:86–94.
55. Hasund A. The Bergen Technique. Norway: University of Bergen, 1972:49.
56. Langlade M. Thérapeutique orthodontique, ed 3. Paris: Maloine, 1986.
57. Samuels RHA, Rudge SJ, Mair LH. A comparison of the rate of space closure using a nickel-titanium spring and an elastic module: A clinical study. Am J Orthod Dentofacial Orthop 1993;103:464–467.

58. Sonis AL, Van der Plas E, Gianelly A. A comparison of elastomeric auxiliaries versus elastic thread on premolar extraction site closure: An in vivo study. Am J Orthod 1986;89:73–78.
59. Kuster R, Ingervall B, Burgin W. Laboratory and intra-oral tests of the degradation of elastic chains. Eur J Orthod 1986;8:202–208.
60. Barrett RD, Bishara SE, Quinn JK. Biodegradation of orthodontic appliances, Part I: Biodegradation of nickel and chromium in vitro. Am J Orthod Dentofacial Orthop 1993;103:8–14.
61. Maijer R, Smith DC. Biodegradation of the orthodontic bracket system. Am J Orthod Dentofacial Orthop 1986;90:195–198.
62. Edman B, Möller H. Trends and forecasts for standard allergens in a 12-year patch test material. Contact Dermatitis 1982;8:95–104.
63. Bishara SE, Barrett RD, Selim MI. Biodegradation of orthodontic appliances, Part II: Changes in the blood level of nickel. Am J Orthod Dentofacial Orthop 1993;103:115–119.
64. Kusy RP. Materials and appliances in orthodontics: Brackets, arch wires and friction. Curr Opin Dent 1991;1:634–644.
65. Scott GE Jr. Fracture toughness and surface cracks–The key to understanding ceramic brackets. Angle Orthod 1988;58:5–8.
66. Kusy RP. Morphology of polycrystalline alumina brackets and its relationship to fracture toughness and strength. Angle Orthod 1988;58:197–203.
67. Akgündüz O. Study of the Structural Characteristics of Ceramic Brackets [thesis]. İzmir, Turkey: Aegean University, 1999.
68. Vaughan JL, Duncanson MG Jr, Nanda RS, Currier GF. Relative kinetic frictional forces between sintered stainless steel brackets and orthodontic wires. Am J Orthod Dentofacial Orthop 1995;107:20–27.
69. Angolkar PV, Kapila S, Duncanson MG Jr, Nanda RS. Evaluation of friction between ceramic brackets and orthodontic wires of four alloys. Am J Orthod Dentofacial Orthop 1990;98:499–506.
70. Bazakidou E, Nanda RS, Duncanson MG Jr, Sinha P. Evaluation of frictional resisatance in esthetic brackets. Am J Orthod Dentofacial Orthop 1997;112:138–144.
71. Taneja P, Duncanson MG Jr, Khajotia SS, Nanda RS. Deactivation force-deflection behavior of multistranded stainless steel wires. Am J Orthod Dentofacial Orthop 2003;124:61–68.

Analysis of Two-Tooth Mechanics

This chapter details the force systems created by straight and bent wires placed in the brackets of two-tooth segments. In a crowding case, to align teeth positioned at different angles and levels, an elastic, straight archwire in the brackets is engaged. The teeth move in response to the wire's elasticity. This creates a force system formed by one wire and multiple brackets. Over time, this force system will reach static equilibrium. The first condition needed for this to occur is for the sum of the vectors of all the forces and moments on the system to be zero. However, to decipher all the forces and moments in a complete dental arch is extremely complex because of the differing anchorage values of multiple teeth, each having brackets and tubes of different widths and being located at various angulations and levels to each other. Therefore, for simplicity, this chapter considers two-tooth mechanics, as suggested by Burstone and Koenig.[1]

Statically Determinate Force Systems

If the distance between two attachments is known and if the force applied by the wire on the bracket can be measured with a dynamometer, the final positions of the teeth can be predetermined. In other words, the moment-and-force system can be controlled in order to guide the tooth movement. This is called a *statically determinate force system*.

For example, Fig 3-1a shows a canine and a premolar with equal anchorage values. A round, flexible straight wire is placed in the premolar bracket and attached to the canine bracket. If the interbracket distance is 7 mm, and the force needed to activate the wire up to the canine level is 100 g, a counterclockwise moment of 700 g-mm should be applied to the premolar bracket. Because the anchorages are equal, the center of resistance of this force system is at the halfway point. For the system to reach static balance, the algebraic sum of all the moments and forces acting on the system must equal zero. The counterclockwise moment on the premolar must be balanced by the clockwise moment created by the equal and opposite vertical forces acting on both teeth. Therefore, when the wire is engaged in the canine bracket, a counterclockwise moment of 700 g-mm on the premolar and balancing vertical forces of 100 g on both sides occur. When the wire is totally deactivated and the teeth reach a statically balanced position, the premolar tips mesially and intrudes while the canine extrudes. However, the canine cannot reach the occlusal plane (Fig 3-1b).

3 | Analysis of Two-Tooth Mechanics

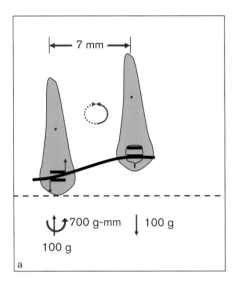 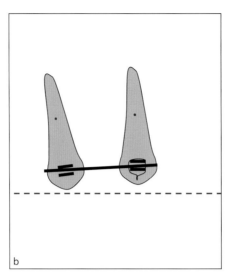

Fig 3-1 Two teeth, a premolar and a high canine with equal anchorage values. *(a)* When a section of wire is placed in the premolar bracket slot and is activated by ligating to the canine, a force system consisting of two teeth and a wire is created. This force system reaches static equilibrium by intrusion (ie, an upward balancing force) and a slight counterclockwise tipping of the premolar as well as extrusion (ie, downward balancing force) of the canine. *(b)* Note that the canine cannot reach the occlusal plane because there is some intrusion of the premolar and the anchorage values are the same for both teeth.

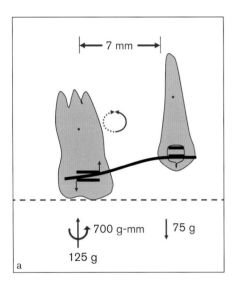 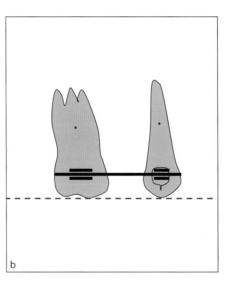

Fig 3-2 Two teeth with different anchorage values. *(a)* If a segmented wire is placed in the molar tube first, then deflected up to the canine and tied at a single point as shown in Fig 3-1, then a counterclockwise moment occurs on the molar. This moment reaches static equilibrium by the vertical balancing forces tending to rotate the system clockwise. These balancing forces are intrusive on the molar and extrusive on the canine. *(b)* Note that the canine reaches the occlusal plane because of the heavier molar anchorage. The molar resists tipping because the anchorage is large enough to resist any movement.

If the premolar in the previous example is replaced with a molar having a higher anchorage value than the canine (Fig 3-2a), then the force system will be different. In this case, a 700 g-mm counterclockwise moment occurs on the molar. This moment is balanced by a 125-g upward force on the molar and a 75-g downward force on the canine. This changes the force system in two ways. First, according to lever principles, as the center of resistance of the system is closer to the molar, the magnitude of the balancing force on the molar is higher than that of the canine. Second, from a clinical point of view, because the molar's anchorage is much higher than the canine's, it does not move. On the other hand, although the magnitude of force is low, extrusion of the canine to the occlusal plane occurs easily (Fig 3-2b).

The relationship between crowded teeth is often complex, which makes it difficult to predetermine tooth movement. Axial inclinations and bracket positions of crowded teeth vary greatly. The previous examples were concerned only with the anteroposterior relationships of two teeth on the same plane, and they did not involve the rotations or angulations of related brackets. When a wire is engaged in two brackets, it affects the teeth in three planes of space (ie, sagittal, transverse, and vertical).

Fig 3-3 When a wire is placed in the brackets of two teeth that have the same anchorage values and are on the same plane, these teeth will not move because the wire does not apply any force on the teeth.

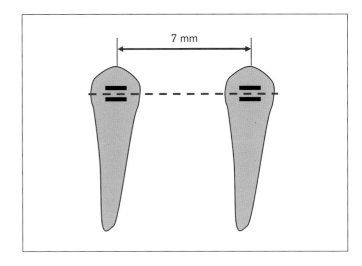

Burstone geometry classes

As the following examples illustrate, different ratios between bracket angulations affect tooth movements. Burstone and Koenig[1] named six geometry classes that can occur between two teeth (for simplification, only the deactivation force systems are shown here). As a starting point, Fig 3-3 shows a 0.016-inch straight wire in two brackets of the same width on the same plane with an interbracket distance of 7 mm. Assuming that the anchorages of the teeth are equal and the wire is not active, no movement occurs. In Fig 3-4, each of the six geometry classes proposed by Burstone and Koenig[1] is illustrated as the angulation of the left bracket is changed while the right one is kept stationary. Figure 3-5 shows the shape a straight wire takes when it is placed in the bracket slots for each of the geometry classes.

Class I geometry

In Class I geometry, the ratio between bracket angulations is $\theta A/\theta B = 1$. Therefore, the angulations of the brackets are equal and their direction is the same. Since the wire-bracket angulations are equal, clockwise moments of 1,860 g·mm occur on each bracket. Because the moments are equal, the ratio between the moments is $MA/MB = 1$. To reach a statically balanced position, the sum of all the moments and forces acting on the system must be zero. In this system, clockwise moments (1,860 g·mm) are acting on both brackets; thus, the sum of the clockwise moments is 3,720 g·mm. For the system to be statically balanced, net forces of 531.4 g occur in an upward direction on the right and in a downward direction on the left. According to the $L \times F$ formula, the system becomes balanced with a counterclockwise moment of $7 \times 531.4 = 3,719.8$ g·mm. From the clinical point of view, when such a force system reaches a statically balanced position, both teeth make clockwise rotations, the right one extruding and the left one intruding.

Class II geometry

In Class II geometry, the ratio between bracket angulations is $\theta A/\theta B = 0.5$, and the ratio between moments is $MA/MB = 0.8$. Since the angulation of the bracket on the right does not change, the amount and direction of the moment acting on this bracket remain the same (ie, 1,860 g·mm in the clockwise direction). Because the wire-bracket angulation on the left bracket is lower than the other, the direction of the moment remains the same, but its magnitude is reduced to 1,488 g·mm. The sum of the moments in the clockwise direction affecting the system is 3,348 g·mm. To balance this, the amount of balancing force for each bracket must be $F = 3,348/7 = 478.2$ (although researchers[1] found a force of 477.4 g to be effective). Because the anchorage values of the teeth are assumed to be equal, the magnitude of the balancing forces on both sides is also equal. This system is balanced with clockwise rotations on both sides; the right tooth extrudes and the left one intrudes.

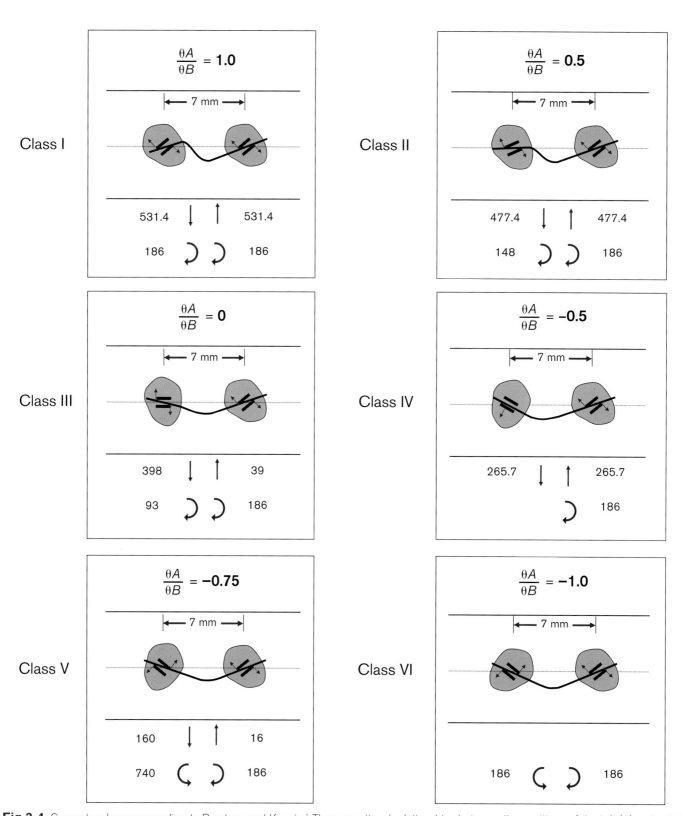

Fig 3-4 Geometry classes according to Burstone and Koenig.[1] The proportional relationships between the positions of the left (A) and right (B) brackets are shown in the top rows, the ratios between left and right moments are shown in the center rows, and the direction and magnitude of the moments and forces (deactivation) that occur on the left and right brackets are shown in the bottom rows. Note that in Class IV geometry, the wire creates a moment of 1,860 g-mm as it enters the right bracket at an angle, whereas it enters the left bracket passively at no angle. (Reprinted from Burstone and Koenig[1] with permission.)

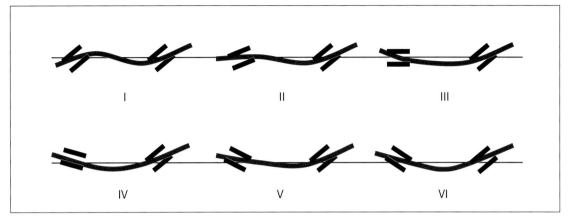

Fig 3-5 The shape of the wire after being placed in the brackets in the six classes of bracket angulation as explained in Fig 3-4. Note that in Class I and Class II geometry, the wire passes over the interbracket plane, while in the other relationships, it remains below this plane. In Class VI geometry, the proportional relationship between the brackets is −1.0. As the brackets are placed at equal and opposite angles, the shape of the wire becomes symmetric; therefore, the moments on both sides are equal and opposite.

Class III geometry

In Class III geometry, the ratio between bracket angulations is $\theta A/\theta B = 0$, and the ratio between moments is $MA/MB = 0.5$. The same results as the previous examples will be achieved because the moment on the right bracket is in a clockwise direction with a magnitude of 1,860 g-mm; the moment on the left bracket is in the same direction, but the wire-bracket angulation is less (930 g-mm). The total clockwise moment in the system (2,790 g-mm) is balanced with a net force of 398 g that rotates the system counterclockwise. When the system reaches a statically balanced position, both teeth rotate in the clockwise direction and the right tooth extrudes while the left one intrudes.

Class IV geometry

In Class IV geometry, the ratio between bracket angulations is $\theta A/\theta B = -0.5$, and the ratio between moments is $MA/MB = 0$. Note that there is no moment acting on the left bracket. This means that the wire enters the left bracket slot without any angulation. In this case, there is only a clockwise moment of 1,860 g-mm in the system. When the system reaches a statically balanced position, the right tooth rotates in the clockwise direction and extrudes with 265.7 g, while the left tooth remains straight and intrudes with the same amount of force.

Class V geometry

In Class V geometry, the ratio between bracket angulations is $\theta A/\theta B = -0.75$, and the ratio between moments is $MA/MB = -0.4$. In this relationship, there is a clockwise moment of 1,860 g-mm on the right and a counterclockwise moment of 740 g-mm on the left. For the system to be statically balanced, a counterclockwise moment of $1,860 - 740 = 1,120$ g-mm is needed. Therefore, an upward force of $1,120/7 = 160$ g on the right and a downward force of the same magnitude on the left occurs. As the system becomes balanced, the tooth on the right rotates clockwise and extrudes while the one on the left rotates counterclockwise and intrudes.

Class VI geometry

In Class VI geometry, the ratio between bracket angulations is $\theta A/\theta B = -1$, and the ratio between moments is $MA/MB = -1$. Note that the angulations of both brackets are equal, but their directions are opposite. Because the bracket angulations are equal, the entrance angulations of the wire in the brackets are also equal. In this case, equal and opposite moments occur on both sides, and the system is statically balanced. Therefore, there are no balancing forces in this system.

Summary

Teeth can be classified as independent units, each with its own center of resistance. When a tooth moves, the other teeth will also move to some extent. All the teeth in a dental arch ligated to an archwire move with the elasticity of the wire by taking support from each other (called *reciprocal anchorage*). When the archwire reaches a statically balanced position, the occlusal plane will have a certain inclination related to the reference plane (ie, the cranial base). The inclination of the occlusal plane is dependent mainly on the positions of the teeth, the general form of the dental arch, and the axial inclinations of the brackets. Occasionally, this may result in an undesired deep bite or open bite.

Mechanics of V-Bend Arches

During orthodontic treatment, various bends are made in the wire to achieve the desired tooth movements. Tipback, gable, and sweep bends of the second order are V-shaped bends (V-bends). These bends are often used in fixed orthodontic mechanics; understanding their force systems and using them conscientiously enables treatment to be carried out more efficiently.

When a V-bend is made in a wire halfway between two teeth with equal anchorage (eg, premolar and canine) and placed in the premolar bracket, the mesial end of the wire will lie gingival to the canine bracket (Fig 3-6a). In this case, the wire is passive (no force system); therefore, no movement is expected. When the mesial end of the wire is inserted into the canine bracket slot, a force system consisting of two teeth and an archwire is created (Fig 3-6b). Because the anchorages are equal, the center of resistance of the system is located at the midpoint of the centers of resistance of the teeth.

This V-bend archwire delivers a clockwise moment to the premolar bracket and a counterclockwise moment to the canine bracket. The sum of all forces acting on the force system and the sum of all the moments around a point must equal zero (see the previous section on statically determinate force systems). With the effect of equal and opposite moments, the teeth will tip around their centers of resistance in a way that their crowns will move away from each other (Fig 3-6c).

When the wire becomes totally passive as the force on the brackets drops to zero, the system will be statically balanced. If the crowns are ligated to each other, the roots will move toward each other around the center of rotations located on the crowns. When the wire becomes passive, the axial inclinations of the teeth will also be corrected (Fig 3-6d). This modification, used in the edgewise technique, is called a *gable bend*.[2]

If the V-bend is moved from the midpoint to one-third of the way from the premolar bracket to the canine bracket (Fig 3-7a), while a clockwise moment occurs on the premolar, no moment (couple) occurs on the canine because the wire is passive in the slot. The mesial tip of the wire applies just a single upward force on the canine bracket. The system reaches static equilibrium by extrusion of the premolar and intrusion of the canine. If the V-bend is moved closer to the premolar bracket, the wire-bracket angulation there becomes higher than that at the canine (Fig 3-7b). Therefore, a larger moment occurs on the premolar than on the canine. Note that the directions of the moments are the same, but because the V-bend is moved toward the premolar, the wire direction entering the canine bracket changes as the wire passes over the interbracket reference line. Static equilibrium is reached here with clockwise rotations on both sides and intrusion of the canine and extrusion of the premolar. Note that the magnitudes of the balancing forces are higher than the ones in the previous examples, because there are two moments with the same direction in the system.[3–7]

The next example involves two teeth, a molar and a canine, with different anchorage values. The V-bend in Fig 3-8a is in the middle, so the wire applies equal and opposite moments to both sides. In Fig 3-8b, however, the V-bend is at the one-third mark. This results in a clockwise moment and extrusive force on the molar and an intrusive force on the canine. Because the center of resistance of this system is close to the molar, the magnitude of the extrusive force is higher than the intrusive force on the canine. In Fig 3-8c, the V-bend is closer to the molar tube, which results in a higher moment on the molar. The direction of the moments is the same because the wire passes over the interbracket reference line; therefore, the balancing forces (extrusion on the molar, intrusion on the incisor) are higher than in the previous examples. Clinically, extrusion occurs

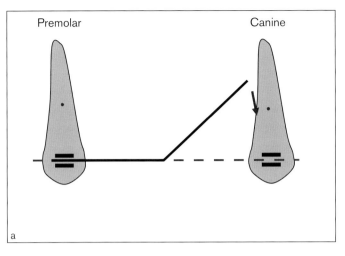
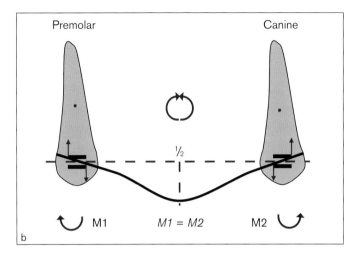
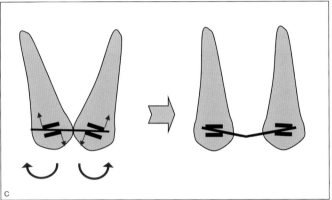
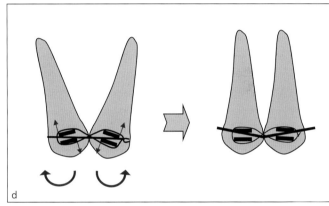

Fig 3-6 *(a and b)* When a V-bend is made halfway between the brackets of two teeth with the same anchorages, the wire enters in the brackets at the same angle after being activated. Therefore, the moments will be equal and opposite on both sides. *(c)* A V-bend placed at the halfway point of two teeth whose crowns have tipped toward each other helps upright the roots while the crowns separate from one another. *(d)* If the crowns are tied together rigidly, the roots become upright while the crowns remain stationary. This is referred to as *gable bend* mechanics and is used in the edgewise technique.

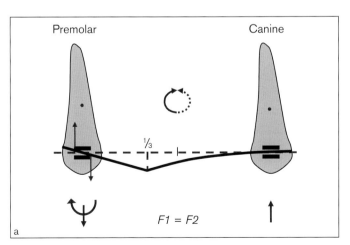
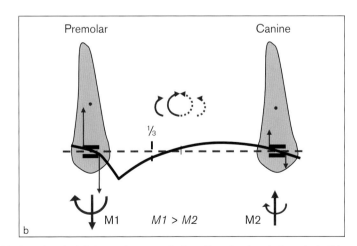

Fig 3-7 *(a)* The force system changes if the V-bend moves to one-third the interbracket distance because it alters the wire-bracket angle. In this configuration, the wire applies a clockwise moment to the premolar bracket as it enters the canine bracket passively. This creates a downward balancing force on the premolar and an upward balancing force on the canine. *(b)* A new force system occurs when the V-bend moves closer to the premolar bracket. The wire-bracket angle on the premolar (ie, the moment) is higher than that on the canine. In this configuration, the wire passes over the interbracket reference plane; thus, the direction of the moment on the premolar is the same as that on the canine.

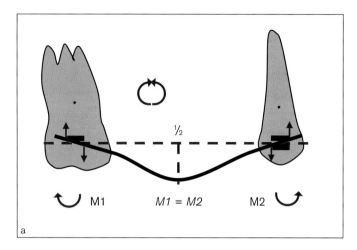

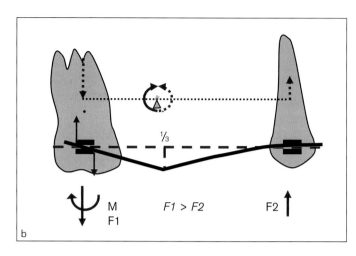

Fig 3-8 *(a to c)* If V-bend arch mechanics similar to those in Figs 3-6 and 3-7 are applied between two teeth of different anchorage values, a different force system occurs. The difference here is the magnitudes of the balancing vertical forces. As the center of resistance of the system is closer to the molar (higher anchorage), the magnitude of the balancing force on this tooth will be higher than that on the canine.

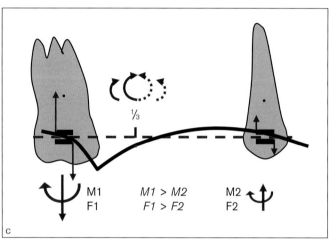

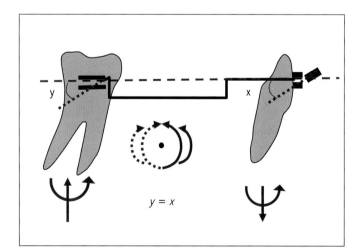

Fig 3-9 When a utility (V-bend) arch having equal tipback and labial root torque angles is placed in the brackets, the extrusive effect on the molars will be higher than the intrusive effect on the incisors because the center of resistance is close to the molars. (x and y represent torque and tipback angles, respectively.)

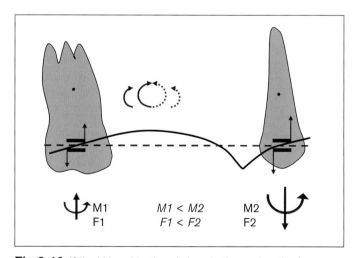

Fig 3-10 If the V-bend is placed close to the canine, the force system created is balanced with intrusion on the molar and extrusion on the canine. Clinically, however, extrusion of the canine occurs much more readily than intrusion of the molar.

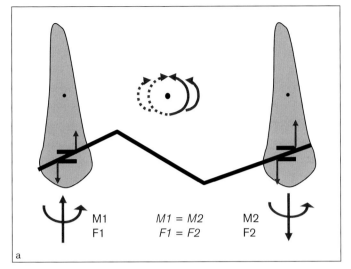 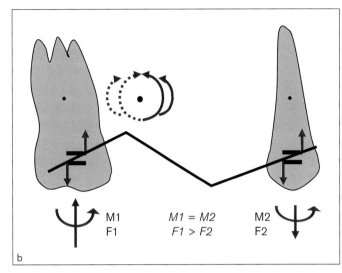

Fig 3-11 *(a)* When a stepped arch is placed in the brackets of two teeth with equal anchorage values, equal and same-directional moments occur on both sides. *(b)* If the anchorages are different, the balancing forces will be of different magnitudes because the center of resistance of the force system is closer to the higher anchorage (molar) side.

easily and causes the bite to open. A utility (2 × 4) archwire with labial root torque on the incisors and tipback on the molars is a good example of this system (Fig 3-9). In this situation, counterclockwise moments occur on both sides. As the center of resistance of the system is closer to the molar, the magnitude of the balancing forces is higher (in some instances, it may be practical to show the center of resistance of a 2 × 4 archwire at the midpoint of anterior and posterior teeth).

If the V-bend is moved closer to the canine bracket (Fig 3-10), the moment on the canine becomes higher than that on the molar. This results in intrusion of the molar and extrusion of the canine. Clinically, intrusion of the molar is difficult because of its high anchorage; therefore, the net effect would be extrusion of the canine only.

Stepped-Arch Mechanics

Artistic, step-up, step-down, and anchorage bends in Tweed mechanics are examples of step bends. In these mechanics, the moments on both sides are equal and same-directional (ie, $M1/M2 = 1$; Fig 3-11). In stepped arches, changing the location of a step bend between brackets does not affect the force system. Thus, the location of the step bend does not affect the ratio between the moments. Likewise, if the height of the step changes, a linear relationship occurs between the moments; ie, the height of the step bend does not affect the ratio between moments.

In stepped-arch mechanics, the interbracket distance is proportional to the M/F ratio; as this distance increases, the M/F ratio also increases. Stepped-arch mechanics between two teeth with equal anchorage is the equivalent of the mechanics in Class I geometry (see Figs 3-4 and 3-5). In practice, increasing interbracket distance is possible only by using narrow brackets. When the interbracket distance is constant, as the height of the step increases, moments on both sides also increase. High moments mean that the balancing forces will also be that much higher.[3,5]

Artistic bends

Artistic bends are a combination of V-bends and step bends applied to the incisor region in the edgewise technique with the purpose of correcting the second-order axial inclinations of incisors. As the crowns are tipped toward each other with reciprocal anchorage, the roots diverge from one another. Figure 3-12 shows a V-bend between the central incisors. To obtain equal and opposite moments, the V-bend must be located equidistant from each tooth.[3,8]

3 | Analysis of Two-Tooth Mechanics

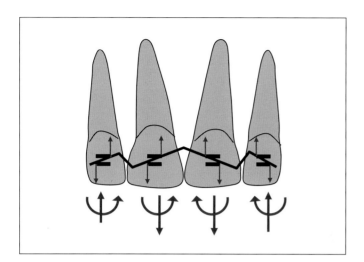

Fig 3-12 Artistic bends are examples of stepped-arch mechanics.

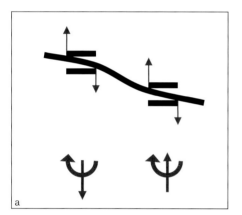 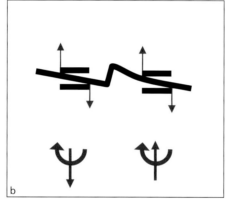 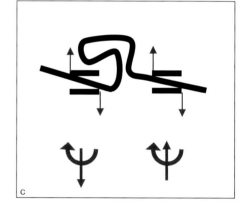

Fig 3-13 A straight arch (*a*), stepped arch (*b*), and looped arch (*c*) between two brackets will give the same mechanical results. In all three relationships, equal and same-directional moments are obtained on both brackets.

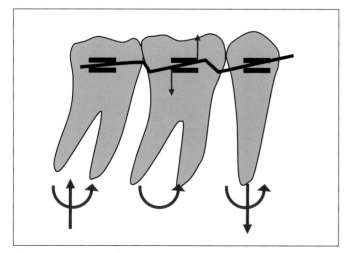

Fig 3-14 Tweed anchorage reinforcement bends are good examples of stepped-arch mechanics.

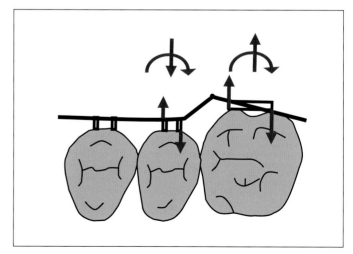

Fig 3-15 Molar offset and toe-in in the first order are also types of stepped-arch mechanics.

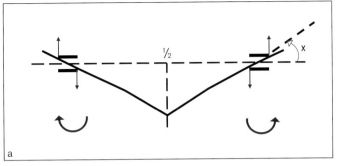 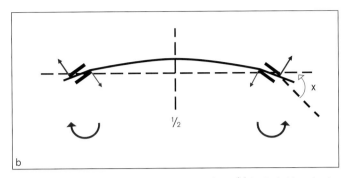

Fig 3-16 *(a)* The V-bend arch located at the center of the interbracket distance is placed in both brackets with an angle x. *(b)* A straight archwire is placed in brackets that are positioned with the same angle x. From a biomechanical point of view, there is no difference between these two applications. The results obtained are absolutely the same, except that in the first example the wire is placed with a bend, and in the second it is placed straight. This shows the difference between the standard edgewise and straight-wire systems.

Step-up and step-down

Step-up and step-down bends are usually used to correct the differences in level between two groups of teeth or to compensate for bracket-positioning errors. Fig 3-13a shows the force system obtained by a straight wire. When a straight wire is placed in the bracket slots, equal and same-directional moments are obtained on both brackets, creating equal and opposite balancing forces. The same mechanical results are obtained with stepped (Fig 3-13b) and looped (Fig 3-13c) arches.

Tweed anchorage bends

In Tweed mechanics, anchorage bends are used to tip mandibular molar crowns distally to reinforce anchorage against Class II elastic forces. With the help of step bends mesial and distal to the mandibular first molars, equal and same-directional moments are obtained on all the teeth (Fig 3-14). This force system reaches static equilibrium when the mandibular second molars have extruded and the premolars have intruded, which may cause bite opening. This is especially important in patients with a vertical facial-growth pattern.[8]

Molar offset and toe-in are examples of step bends in the transverse direction. Figure 3-15 shows the force system that occurs when an arch with a molar offset and toe-in is placed in the attachments: same-directional moments on the molar and second premolar. The balancing forces tend to move the molar buccally and the second premolar palatally.

Straight-Wire Mechanics

In the first mechanism explained previously, a preformed continuous wire is inserted into the preangulated bracket slots. In the second (V-bends and stepped-arch mechanics), an archwire is placed in the bracket slots located on the same plane.[9] In both mechanisms, the moments acting on the brackets are proportional to the wire-bracket angulations and are equal for a certain wire-bracket angulation. In other words, if an interaction between the wire and bracket is desired, it does not matter whether the bracket is angulated to the wire plane or the wire is placed in angulated brackets. If the wire-bracket angulation is the same, the moments applied to both brackets will also be the same (Fig 3-16).

This idea puts forward the main difference between straight-wire and standard edgewise techniques. In the latter, the wire must be bent and actively placed in a nonangulated slot to obtain a moment such as tipback, antirotation, toe-in, antitip, torque, or gable bend. In the straight-wire system, however, the archwire creates an immediate moment on the teeth from the time it is engaged in the preangulated bracket slots. Clinically, the differences between these two configurations have certain mechanical implications.

For example, if the treatment goal is to distalize a canine using a segmented arch in the standard edgewise technique (Fig 3-17a), an antitip bend (13 degrees in the example) must be placed in the wire to achieve the desired controlled tipping. Assuming that the antitip bend applies a counterclockwise moment of 1,050 g-mm

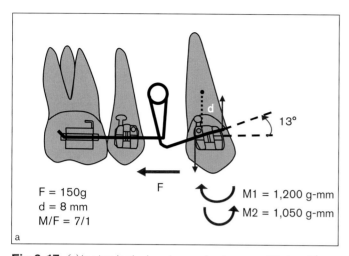 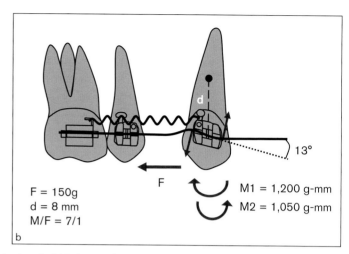

Fig 3-17 *(a)* In standard edgewise mechanics, an antitip bend (second-order bend of 13 degrees) must be placed to move the canine with a controlled tipping movement. *(b)* In straight-wire mechanics, second-order angulation already exists in the bracket, so the moment (M2) starts to work as soon as the straight wire is placed in the bracket.

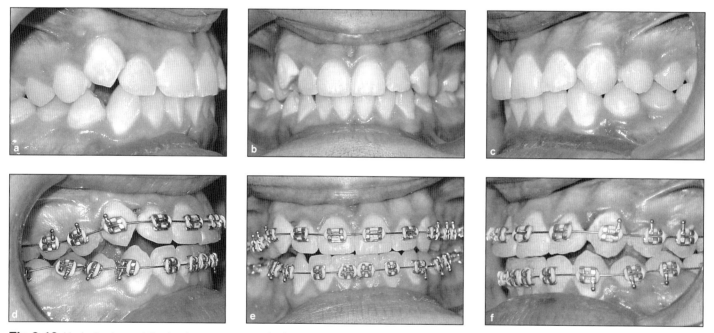

Fig 3-18 Undesired open bite due to engagement of a flexible straight wire in the bracket of a high canine. Initial interdigitation of the posterior segment between premolars *(a to c)* has been lost after only 1 month of treatment *(d to f)*. Note that the open bite occurred mainly because of protrusion of the maxillary incisors and intrusion of the premolars. If the first premolars' had been extracted, the adjacent teeth would have been expected to tip toward the extraction sites.

(M/F ratio, 7:1), a distal force of 150 g must be applied by activating the reverse closing loop. To increase this M/F ratio, either the degree of the antitip bend can be increased or the magnitude of the force can be decreased.

On the other hand, when a straight wire is placed in a straight-wire bracket having a 13-degree angulation, the same 1,050-g-mm positive moment occurs (Fig 3-17b). To distalize this tooth with controlled tipping, making an extra bend is not necessary because the antitip angulation is already built into the bracket. With this moment, the crown tends to move mesially and the root distally. To reach the M/F ratio of 7:1, applying a distal

Fig 3-19 Elastic archwire inserted into a high canine bracket deforms the general arch form and sometimes causes canting of the occlusal plane.

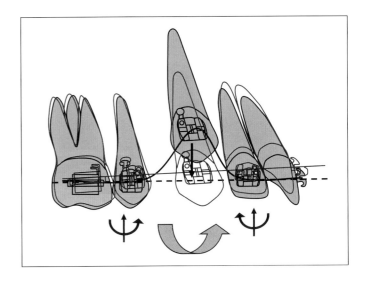

force of 150 g to the bracket will be sufficient. In this example, friction between the wire and bracket has not been taken into consideration.

Straight-wire mechanics depend on bracket inclinations, tooth positions, and general form of the dental arch. It is usually difficult to predetermine what will be the final cant of the occlusal plane when a straight wire is passed through all the brackets. In other words, straight-wire mechanics are statically indeterminate. Figure 3-18 shows an example of an adverse effect that can occur with a straight wire placed in the bracket of a high canine. Owing to the position of the canine, an anterior open bite has occurred by the incisors' flaring and intrusion. Elastic archwires placed on high canines (Fig 3-19) or ectopic teeth tend to deform the general arch form and may cause canting of the occlusal plane. Figure 3-20 shows three examples of mechanics used to avoid the adverse effects of excessive wire deflection.

Laceback

The main difference between the two mechanisms explained above is that in an edgewise system, the M/F ratio is controlled with wire bends, whereas in a straight-wire system, it is predetermined by the angulation built into the bracket. Therefore, to reduce the adverse effects of straight wire–bracket angulations, especially at the beginning of leveling, canines and molars must be attached to each other firmly with a laceback ligature tie (Figs 3-21 and 3-22). In this way, protrusion of the canine crown can be prevented as the root moves distally around the center of rotation located on the crown. This movement continues until the wire is completely deactivated. If further distal movement of the canine is desired, the laceback can be activated slightly. This causes the tooth to return from root movement back to controlled tipping.

If the laceback is activated excessively, the canine crown tips farther distally at a rate allowed by the bending stiffness of the wire. During this process, as the angulation between bracket and wire increases, the friction between these materials also increases. This results in binding, which stops tooth movement. To resolve the binding, one must wait until the tooth is totally uprighted so the wire is deactivated. The length of this period depends on the stiffness of the wire and biologic factors such as the length of the root and the age and density of the alveolar bone. Clinically, binding is a serious adverse mechanical effect that can cause loss of anchorage and wasted time. For more on binding, see chapter 4.

If the canines are inclined vertically or are distally positioned, a straight wire passing through the angulated canine bracket slot falls below the incisor brackets.

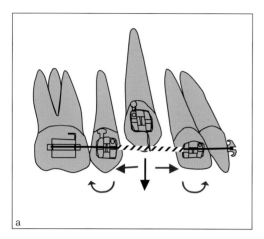
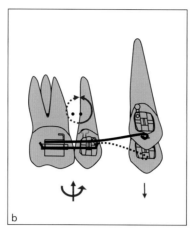

Fig 3-20 Four examples of mechanics used to extrude a canine. (*a*) An open coil spring between the lateral incisor and premolar on 0.016-inch stainless steel wire maintains the space while preventing the adjacent teeth from tipping. (*b*) A cantilever with a V-bend can be used to move the canine down. The cantilever should be attached to the canine with a ligature at only one point to avoid unwanted moment. (*c*) Reciprocal anchorage to level maxillary and mandibular canines with an up-and-down elastic. (*d*) An auxiliary 0.014 or 0.016 NiTi wire can be used along with a rectangular SS main archwire to bring the high canine down.

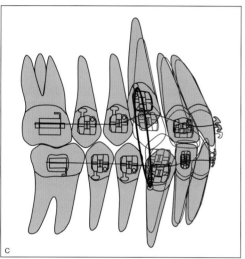
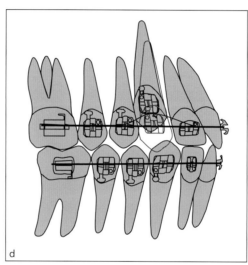

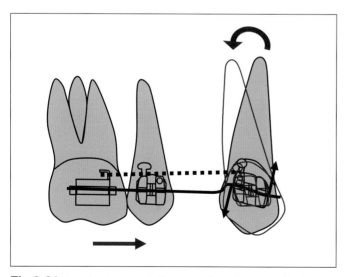

Fig 3-21 Laceback prevents the crown from tipping and helps correct the inclination of an upright canine.

When the wire is engaged in the incisor brackets, three possible effects can be observed.

First is the rowboat effect, which is caused by a counterclockwise moment on the canine that strains the anchorage.[10–12] This moment tends to push the crown forward, resulting in incisor protrusion, which can be prevented only by means of a laceback.

In Class II, division 1 extraction cases, the rowboat effect is an undesirable side effect because of the round tripping or jiggling effect,[13–15] which may occur during retraction of anterior teeth and result in root resorption. Laceback can prevent the canine crown from tipping forward. In Class II, division 2 nonextraction cases, incisor protrusion may be desirable; therefore, use of a straight wire will help induce anterior protrusion as well as quick alignment.

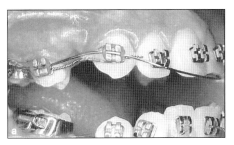 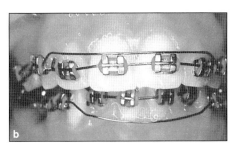 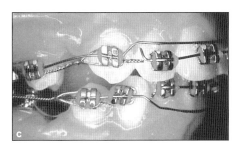

Fig 3-22 *(a)* To avoid extrusion, if a straight wire passing through a vertically positioned canine bracket passes below the incisors, one should not place the wire in the incisor brackets. *(b and c)* In this case, instead of using nickel titanium (NiTi) wires, thin stainless steel wires with step-up bends bypassing the incisor brackets may be used. This is a desirable approach for treating anterior open bites needing correction by incisor extrusion.

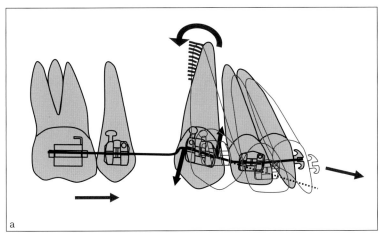 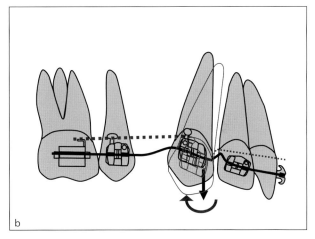

Fig 3-23 *(a)* The bowing effect may cause bite deepening because of the canine position. *(b)* If a bendable wire is used, a step-up can be bent to bypass the incisors gingivally. If the stepped arch is engaged in the anterior brackets, it will cause the canine crown to tip distally because of the clockwise moment.

The second effect is deepening of the bite (bowing effect; Fig 3-23a). Deepening of the bite during treatment is usually not a desired effect unless it is indicated in an anterior open bite that needs to be corrected by maxillary incisor extrusion. To diagnose or predetermine this effect, place the archwire in the canine bracket slot before ligating it. If the anterior part of the wire runs below the incisor brackets, it should not be tied to the brackets to avoid incisor extrusion. If a bendable wire is used, a step-up can be bent to bypass the incisors gingivally. Another method is to place a continuous intrusion arch along with the straight wire. The extrusive effect of the straight wire would therefore be compensated for by the intrusion arch.

If the stepped, bypass archwire is not left passive and is engaged in the incisors to intrude them or prevent them from extruding, it will cause the canine crown to tip more distally owing to the clockwise moment occurring on its bracket (Fig 3-23b).

In the explanations above, the main reasons for the adverse effects are the positions or axial inclinations of the teeth or the brackets. If the problem is caused by the axial inclinations of the canines, it is important to upright them with a laceback before inserting a continuous wire.

Conclusion

In the analysis of the relationship between two teeth, the slot sizes and bracket widths are assumed to be equal in all the examples given here. Naturally, as the slot sizes and widths change, the magnitudes of the balancing forces and the moments also change. In clinical application, the density of the alveolar bone in which the teeth are located; patient's age, occlusal forces, habits, soft tissues, and parafunctions; number and length of the roots; and other factors can affect the biomechanics of tooth movement. These biomechanics depend on basic physical laws, which must be considered in designing archwire configurations and their placement into preadjusted brackets. Straight-wire mechanics are combinations of two-tooth relationships. When an archwire is ligated, it is not possible to accurately predetermine the magnitude of the moments and balancing forces nor their directions, making them statically indeterminate mechanics. Straight-wire mechanics are also called *shape-driven mechanics* because they are directly affected by the locations of the teeth, their axial inclinations, and the arch forms.[8]

References

1. Burstone CJ, Koenig HA. Force systems from an ideal arch. Am J Orthod 1974;65:270–289.
2. Ülgen M. Treatment Principles in Orthodontics, ed 3. Ankara: Ankara University, 1990.
3. Burstone CJ, Koenig HA. Creative wire bending–The force system from step and V bends. Am J Orthod Dentofacial Orthop 1988;93:59–67.
4. Ronay F, Kleinert W, Melsen B, Burstone CJ. Force systems developed by V bends in an elastic orthodontic wire. Am J Orthod Dentofacial Orthop 1989;96:295–301 [erratum 1990;98:19].
5. Demange C. Equilibrium situations in bend force systems. Am J Orthod Dentofacial Orthop 1990;98:333–339.
6. Bequain DMJ. Etude mécanique de la coudure d'un fil. Orthod Fr 1994;65:547–557.
7. Deblock L, Petitpas L, Ray B. Mécanique de recul incisif maxillaire. Orthod Fr 1995;66:667–685.
8. Burstone CJ. The biomechanical rationale of orthodontic treatment. In: Melsen B (ed). Current Controversies in Orthodontics. Chicago: Quintessence, 1991:131–146.
9. Mulligan TF. Common Sense Mechanics. Phoenix, AZ: CSM, 1982.
10. Bolender CJ. Le torque progressif, une innovation gratifiante de la technique Tip-Edge. Orthod Fr 1995;66:901–916.
11. Mulligan TF. Common sense mechanics. J Clin Orthod 1980; 14:180–189.
12. Isaacson RL, Lindauer SJ, Rubenstein LK. Moments with the edgewise appliance: Incisor torque control. Am J Orthod Dentofacial Orthop 1993;103:428–438.
13. Roth RH. Five year clinical evaluation of the Andrews straight-wire appliance. J Clin Orthod 1976;10:836–850.
14. Roth RH. The straight-wire appliance 17 years later. J Clin Orthod 1987;21:632–642.
15. Viazis AD. Bioefficient therapy. J Clin Orthod 1995;24:552–568.

Frictional and Frictionless Systems

Frictional Systems

Friction

When two objects in contact are forced to move on each other, the resistant force that occurs at the contact surface opposite the direction of movement is friction. Friction that exists before one of the objects starts to move is called *static frictional force*. Static friction is the amount of force necessary to start movement of an object in a static state. Kinetic friction (or dynamic friction) is the friction that exists during the movement of the object, and it is the amount of force that the object must overcome to continue moving. A typical friction graph is shown in Fig 4-1. Static friction is proportional to the force; as the force increases, the friction also increases. When the force comes to a critical point (f max), the static friction is overridden and the object starts to move. From this point on, resistance to movement of the object is called *kinetic friction*. Theoretically, kinetic friction has lower values than static friction.

Friction between solid objects can be rolling or sliding, depending on the type of movement. Because orthodontic tooth movement is a slow process, the wire and bracket relationship can exhibit both static and kinetic forms of sliding friction because the application of force starts a complex biomechanical relationship between the wire-bracket-ligature-tooth-periodontium system and the alveolar bone. Before we give this relationship any detailed consideration, it is necessary to understand friction.

When a book on a table is in a stable state, it applies a load or force (A) equal to its weight on the table (Fig 4-2a). The table conducts a force (N) on the book of the same magnitude and in the opposite direction. Force A is the result of many force vectors spread over the entire surface of the book; if the book is of a homogenous structure, these force vectors will spread out equally on the contact surface, so the resultant vector of these forces is located in the geometric center of the book. The same conditions are valid for the resultant force N.

If the book is pushed lightly from left to right (Fig 4-2b), because of the force acting on the book (P), the magnitude, direction, and point of application of the force N begin to change, and so does the book's homogenous distribution. The cause of this is the roughness of the contact surface. Let's assume for a moment that the contact surface is a frictionless and vacuumed environment. In this case, after starting movement with a minor force (P), the book would maintain its velocity

4 | Frictional and Frictionless Systems

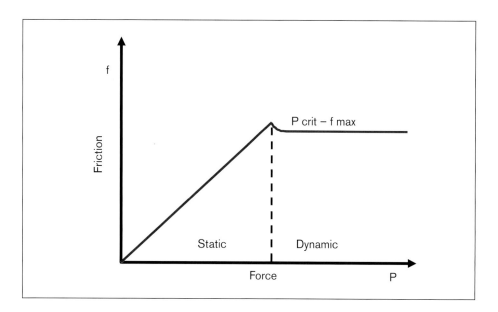

Fig 4-1 A typical friction graph. As the activating force (P) increases, the frictional force (f) also increases. The object starts to move as soon as the static frictional resistance is overridden. Dynamic (kinetic) friction, which is slightly less than static frictional force, starts from this point forward.[1] (Reprinted from Nikolai[1] with permission.)

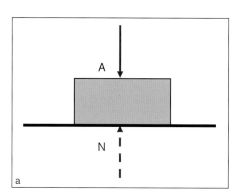 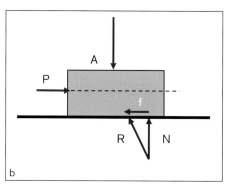

Fig 4-2 *(a)* When a book on a table is in a stable state, it applies a load (A) equal to its weight on the table. The table also applies equal and opposite force (N) on the book. *(b)* If a light horizontal force (P) is applied to the book from left to right, the book does not move immediately because of the static frictional force (f) between the book and the table. This force, which is tangent to the contact surface between the book and the table and opposite in direction, creates a resultant force (R).

and direction in accordance with Newton's first law, and no changes in the magnitude, direction, or point of application of the N force would occur. However, in this example, there is a rough, interlocking contact surface between the table and the book that could be seen if examined under a microscope. When the book is pushed with horizontal force P, it does not start to move immediately because of the resistant force (ie, friction) resulting from the interaction between these two materials. This resistance (frictional force [f]) is equal to and opposite force P, creating a resultant force (R).

If force P is increased up to the critical point, the frictional force f reaches the maximum level (f max). If force P exceeds the critical limit, the static frictional force will be overcome and the book will start to move. The book is now transferred from a static state to a dynamic state, and the frictional force f becomes dynamic (f dynamic) and decreases.

The values of f max and f dynamic depend on the magnitude of force A (the weight of the book in the example) and the coefficient of friction (μ) of the surfaces in contact ($f = \mu \times A$). This means that whether the object is in a static or dynamic state, the frictional force depends on the coefficient of friction of the surfaces in contact and the normal force between objects. In other words, the frictional resistance between the book and the table in this example depends on the weight of the book and the roughness of the surfaces. As the book's normal force (weight) or the coefficient of friction increases, the friction also increases, and the force necessary to initiate the movement of the book (to overcome f max) also increases proportionally. Therefore, there

Fig 4-3 Friction exists during all movements in which the wire is in contact with the bracket and ligature. During leveling, when a flexible wire is placed in a high canine bracket, the tooth moves downward with the elasticity of the wire. As the curved wire is deactivated, it straightens by sliding through the adjacent brackets, with friction occurring between the wire, bracket, and ligature materials.[2]

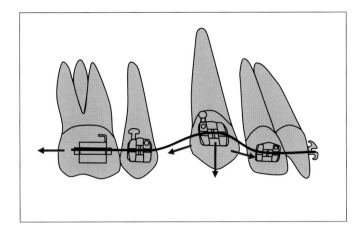

is a direct proportion between force P and force f (ie, friction).

Theoretically, this equation is independent of the width of the contact surface. For example, when we push a box along the floor, the friction is always the same whether we stand the box vertically or horizontally. Whereas, if we push two boxes placed on top of one another, the amount of friction will be twice as much because the weight is double; however, a larger contact surface increases the possibility of interaction between the objects. Thus, even though friction in theory is independent of the objects' contact surface, in practice the size of the contact surface is significant and must be considered.

The ratio between load A and friction (f) is constant ($\mu = f/A$). This ratio is called the *coefficient of static and kinetic (dynamic) friction*, and it is shown as μs and μk, respectively. The coefficient of static friction is higher than that of kinetic friction, and it especially concerns orthodontists because a significant portion of the force applied to the tooth is spent overcoming the static frictional resistance between the wire, ligature, and bracket.

Analysis of the bracket-wire frictional relationship

Generally, friction is thought to occur only when a tooth slides along the archwire. Friction occurs, however, in all situations in which the wire is in contact with the bracket or ligature and in which there is a tendency for movement. For example, during leveling, a flexible archwire is placed in the brackets located at different levels and angulations to each other. The movement of teeth slides the wire through the bracket slots and tubes, resulting in friction between all the materials in contact: brackets, tubes, wire, and ligatures (Fig 4-3). Friction has an important impact on the efficiency of orthodontic mechanics because approximately 40% to 50% of the force used for tooth movement is lost to friction.[2,3] Friction, or binding, which prevents the wire from sliding through the bracket slots, can delay and even halt tooth movement.

For a better understanding of the bracket-wire relationship, let's analyze the movements of a canine sliding along the archwire. Before distalization, a passive relationship exists between the canine bracket and the archwire (Fig 4-4a). When a constant, distal force is applied to the bracket, the tooth tips distally and the mesial wing of the bracket applies a downward force while the distal wing applies an upward force, creating a couple (clockwise moment, M1), as shown in Fig 4-4b. The deflected wire applies an opposite couple of equal magnitude to the wings of the bracket (counterclockwise moment, M2). The deflection of the wire is proportional to its stiffness in the load/deflection rate so that, for a given force, the deflection of stiff wires (such as stainless steel [SS]) is fairly low. As the angle (amount of tipping) between the wire and bracket slot increases, the friction also increases proportionally. When a certain point is reached, the wire resists and no more deflection occurs; thus, the tipping stops. The tooth starts to upright with M2 moment as the normal force is reduced. By over-

4 | Frictional and Frictionless Systems

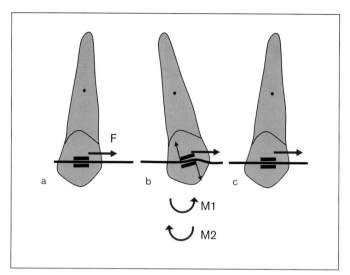

Fig 4-4 Stages of canine distalization with sliding mechanics. From the effect of force F *(a)*, the tooth tips distally (M1) and touches the wire, then starts to upright from the counterclockwise moment (M2) applied by the wire to the bracket *(b)*. When the static frictional resistance between the wire and bracket is overridden *(c)*, the tooth starts to move distally by sliding along the wire. Canine distalization is achieved with multiple tipping-uprighting cycles like this one.

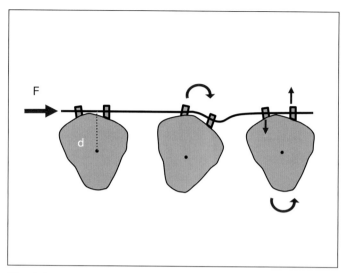

Fig 4-5 In the transverse dimension, distopalatal rotation occurs on the canine because of the force applied to the tooth. This rotational effect can be eliminated by the opposite couple of force that the bracket and ligature apply to the wire. During distalization, several rotational and counterrotational movements occur. d, perpendicular distance from the center of resistance to the line of action of the force (F).

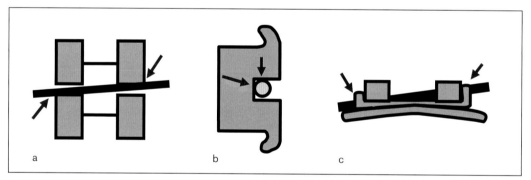

Fig 4-6 *(a to c)* During canine distalization, friction occurs on at least six points between the wire, bracket, and ligature.

coming the static frictional force, the tooth starts to move and slides along the straightened wire (Fig 4-4c).

Excessive tooth tipping (high angulation between archwire and bracket) causes binding, and excessive distal force may lead to permanent deformation of the wire. Canine distalization along a wire is realized with tipping-uprighting cycles.[3]

In the transverse plane, as the force is applied from the buccal side of the canine's center of resistance, the tooth rotates in a distopalatal direction (clockwise rotation; Fig 4-5). This effect is counteracted by an opposite couple from the ligature (counterclockwise rotation). During this distalization, several rotational and counterrotational movements occur.

The amount of tipping and rotation is inversely proportional to the stiffness of the wire. The tooth tips more easily on highly elastic wires; however, uprighting is much harder. On a stiff wire, the tooth tips slightly and

> **Box 4-1 Clinical factors that cause friction**
>
> 1. Bracket
>   - Material[2–8]
>   - Width[3,7,9–15]
>   - Design and manufacturing technique[8,11]
> 2. Archwire
>   - Material[4,6–9,13,16]
>   - Size and cross-section[4,7,10,12,16,17]
> 3. Ligature
>   - Material[3,7,12,17–19]
>   - The use of self-ligating brackets[11,20,21]
>   - Tightness[7]
> 4. Force
>   - Magnitude[22,23]
>   - Point of application[2,24]
> 5. Bracket-wire angulation[3,7,9,12,17,25–27]
> 6. Biologic factors
>   - Saliva[6,7,12,28–31]
>   - Surrounding tissue resistance[3,24]

uprights quickly. The clearance between the wire and bracket slot is also important for the amount of tipping. If the wire is stiff but its diameter is small, the amount of tipping will be that much more. During leveling, thinner wires cause less friction because they can slide easily through the bracket slots. However, in sliding mechanics such as canine distalization, SS wires (0.016-inch or 0.016 × 0.022–inch in 0.018-inch brackets) should be used to keep tipping and friction at an optimum level. In sliding mechanics, a minimum play of 0.002 inch in the bracket slot is necessary to minimize tipping and obtain effective sliding.

During canine distalization, friction occurs on at least six points between the wire, bracket, and ligature (Fig 4-6). Static or kinetic frictional forces can delay or stop tooth movement. Frictional forces are clinically unpredictable because friction is a multifactorial phenomenon. As friction delays tooth movement, the movement site becomes an anchorage site. If the frictional force is excessive, the posterior teeth start to move because the amount of force becomes an optimum force for anchorage teeth.

Factors that affect friction between bracket and wire

Clinically, a number of factors may cause friction (Box 4-1), the amount of which varies depending not only on the magnitude of the force but also on the type of materials used and their surface characteristics. The amount of friction between smooth surfaces is obviously less than that between rough surfaces. However, several factors such as material, tightness of ligature, saliva, bracket width, and wire size can also affect friction. Because the consequences can directly affect the clinical outcome, this subject is examined in detail.

Bracket

Properties related to bracket material Among orthodontic bracket materials, the greatest friction occurs in plastic (polycarbonate)[32,33] and ceramic brackets,[4] and the lowest occurs in SS brackets. To eliminate the drawbacks caused by the friction of plastic—and some ceramic—brackets, manufacturers are inserting a metal slot (eg, Clarity [3M Unitek]) into the ceramic body.[2,5,15,17,26,29,30] Monocrystalline and polycrystalline alumina are two common ceramic bracket material structures. Alumina is known as the third-hardest material,[2] and it has been observed through x-ray element analysis that hard ceramic brackets "scratch" the surfaces of titanium wires.[34] Recently, brackets with smoother slot surfaces have been manufactured to avoid this disadvantage.[2] Even though the surface roughness of monocrystalline alumina is less than that of polycrystalline alumina, their frictional properties are very similar.[28]

Bracket width Some studies claim both narrow[7,10,11] and wide brackets[3,13,14] cause less friction between the wire and bracket. When a tooth tips, the normal force applied to the wire by narrow bracket wings is higher than that of wide brackets (see Fig 1-19); therefore, the friction between the wire and the bracket is expected to be higher. The contradictions between these studies come from differences in the study designs and the materials used. The fact that the wire has more play in the bracket slot of a narrow bracket than in a wide bracket—causing less interaction between the wire and bracket—must not be overlooked. Clinically, medium or wide brackets are preferred, particularly in extraction cases wherein movement control in the transverse plane is important.

In conventional edgewise brackets, the wire is engaged with a ligature by pressing it toward the base of the slot. The firmer this tie, the harder it is for the bracket to slide on the wire, because the normal force is high, causing more friction. Some brackets have been designed and manufactured to avoid this drawback. Friction Free (American Orthodontics) and Synergy (RMO) brackets have been claimed to reduce the normal force of ligation and to allow the wire to move freely in the bracket slot. The frictional resistance values of the Friction Free brackets were found to be much lower than those of other brackets.[8] Nevertheless, these brackets are not fully effective clinically because they do not permit the ligated wire to transfer its full potential to tooth movement.

Bracket manufacturing technique Friction from milled brackets was found to be higher than that from cast and sintered brackets.[35,36] These results were supported by scanning electron microscope views of the bracket slots studied; these pictures showed that sintered brackets have fairly smooth slot surfaces. The edges of cast bracket slots are rougher than those of sintered ones. Milled brackets, however, occasionally have sharp burs on the edges of their slots that may affect frictional resistance.

Wire

Lubrication by saliva The effect of saliva and its role as a lubricant for reducing the amount of friction is controversial. Andreasen and Quevedo[12] used human saliva in their study on frictional resistance and found no difference with or without saliva. They stated that the role of saliva was insignificant. Kusy et al[29] tested the use of human saliva in their experiments on frictional resistance. They reported that saliva only decreased friction with β-titanium and nickel-titanium (NiTi) archwires. The levels registered for SS and chrome-cobalt wires were higher than those obtained in dry state.

Wire material The most commonly used wires, listed from lowest to highest in terms of their surface roughness, are SS, chrome-cobalt, NiTi, and β-titanium. Theoretically, wire-bracket friction increases as the surface roughness of the wire increases. However, this has not been verified either experimentally or clinically.[3,10,13,16] Clinicians know that a NiTi wire, despite its high surface roughness, can "creep" through the bracket slots and tubes, causing soft tissue irritation. This is probably because of its high flexibility and the fact that it moves easily from the effects of chewing cycles and brushing. In addition, saliva may also help reduce the friction by lubricating contact surfaces.

Properties related to wire size and cross-section It is well documented that for the same bracket and wire material, frictional forces increase as the wire size increases, and rectangular wires cause higher friction than round wires do.[4,7,8,12] These values are higher for rectangular NiTi and β-titanium wires, which have rougher surfaces than SS and chrome-cobalt wires.

Factors related to interbracket distance As the interbracket distance increases, stiffness of the wire decreases. For a given force, an elastic wire deflects more than a stiffer one. The more a wire deflects, the more the tooth tips; thus, the angle between the wire and the bracket increases, resulting in binding and higher friction. Increasing the stiffness of the wire to avoid friction might be a solution, but, as mentioned previously, thicker, rectangular wires cause more friction than thinner, round ones. Sometimes high friction can be avoided during canine retraction if the ends of a thin wire are cinched back tightly, stiffening the wire. Clinically, as the clearance between the bracket slot and wire increases, control over tooth movement decreases. Proffit[37] suggests keeping at least 0.002 inch of play between wire size and slot size to sufficiently control movement. This is also true from the friction point of view. According to Drescher et al,[3] the bracket of a canine sliding over a 0.016 × 0.022–inch wire contacts it on the second order (ie, the 0.016-inch side), and both this wire and a 0.016-inch round wire produce less friction than thicker wires. Therefore, 0.016-inch and 0.016 × 0.022–inch SS wires (in 0.018-inch slots) seem to be the best options for controlling both friction and tooth movement.

Ligature

Effect of ligation Archwires are ligated to brackets by three different methods:

- Elastic modules
- Wire ligatures
- Self-ligating brackets

The tightness of the ligature and the wire's pressure on the bracket slot (normal force) directly affect the amount of friction between these materials. In self-ligating brackets, a springy cap or clips join the wire and bracket. The normal force applied by the cap on the wire is standard. Frictional forces obtained with these brackets are approximately the same as[14] or slightly lower than those of elastic or wire ligatures[20] (Fig 4-7).

Wire ligatures cause less friction than elastic ones. However, ligation force (normal force) plays an important role in friction. If the archwire is tightly ligated, there will be high normal force, which will increase resistance to sliding. Elastic ligatures may be better than wire ones because they apply a standard and more controlled force throughout treatment. Practically, because the frictional coefficient does not change, loosening the elastic ligatures before placement or allowing them to relax in the oral environment would not affect the frictional force very much. During sliding mechanics, especially in canine distalization, where friction is a foremost concern, wire ligature is generally preferred over elastic. In this case, it is better to tie the ligature only from the distal wing of the bracket and to tie it loose enough to allow tooth movement. Clinically, it is not easy to standardize the tightness of ligatures. A practical method for this is to tighten the wire by placing the tip of a probe between the ligature and the bracket and moving the ligature slightly after removing the probe. By so doing, the ligature is tight enough to control tooth movement yet loose enough to allow sliding.

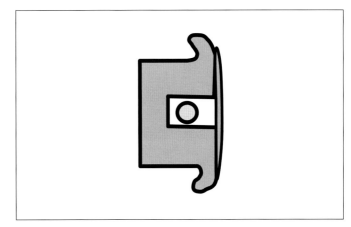

Fig 4-7 Self-ligating brackets allow the wire to move freely through the slot. More play between wire and slot results in less friction but also less movement control.

Force

As explained previously, the main reason for friction during canine distalization is the force applied to the bracket. Because the line of action of this force does not pass through the center of resistance, the tooth tips distally and causes friction between bracket and wire. By reducing the amount of tipping (reducing the force), friction can be reduced. It is therefore necessary to move the line of action of the force closer to the center of resistance. To do this, the force can be applied to a power hook rather than at bracket level.[2] Power hooks have different lengths. Even though this application can be helpful to minimize tipping, long hooks are not practical because they may disturb the gums. They might also cause hygiene problems, because food easily gets stuck underneath. In practice, medium-length hooks (or Kobayashi ligatures) are preferred.

Wire-bracket angle

The wire-bracket angle affects friction significantly: the higher the angle, the greater the friction. This is particularly important when working with preadjusted brackets. A flexible wire engaged in excessively tipped brackets may cause high moments and thus high friction or binding. Binding can be considered a lock preventing the desired tooth from moving and causing anchorage teeth to move instead (ie, anchorage loss). Even though

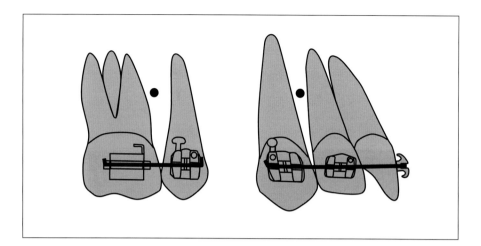

Fig 4-8 In the segmented arch technique, anterior and posterior teeth are incorporated into two segments with rectangular wires. These segments can be considered two big "teeth" having their own centers of resistance.

thin, flexible wires are used at the beginning of leveling, some anchorage loss can be expected. Thanks to new developments in material technology and techniques, today's orthodontists are able to create alternative mechanisms to carry out the desired tooth movement. For example, frictionless mechanics can be used as an alternative to frictional systems.

Frictionless Systems

Because the frictional system is statistically unpredictable and multifactorial, researchers have been pushed to look for more predictable force systems. The segmented arch technique eliminates the disadvantages of the frictional system.

Philosophy of the segmented arch technique

In the segmented arch technique, the dental arch is split into two segments. The anterior segment consists of the incisors and canines, and the posterior segment contains the premolars and molars. The right and left sides of the posterior segment are connected with a stabilizing transpalatal arch. Both segments thus become two big "teeth" with stiff rectangular archwires. In this technique, all movements are converted into a format of relationships between two teeth (as explained in chapter 3), thereby simplifying the force system and providing a predetermined and more controlled mechanism. Each of these two big teeth has its own center of resistance (Fig 4-8).

A force system is predictable or statically determinate if the amount of forces and moments is measurable (see chapter 3). In the segmented arch technique, the magnitude of a force applied to a segment can be measured with a dynamometer; thus, the moment can be easily calculated by measuring the distance between two attachments. Presently, precalibrated wires are commonly used to make results more predictable.[38,39]

In the segmented arch technique, long interbracket distances make it possible to apply light and long-range forces. One can enhance this advantage by using bendable, flexible wires with high working range and low stiffness, such as titanium-molybdenum alloy (TMA).[40,41]

Comparing Frictional and Frictionless Systems in Clinical Orthodontics

Space closure mechanics can be categorized as frictional and frictionless systems. In the frictional system, the teeth move by sliding along the archwire. This can be likened to the movement of a train on rails. In the frictionless system, however, the teeth are carried with the arch by means of loops, which is analogous to a train car being picked up and moved by a crane.

Frictional mechanics are performed on a continuous archwire, including all teeth from molar to molar. Thanks to preformed straight wires with very low load/deflection rates and high springback, such as NiTi, leveling is no longer a problem. Furthermore, chair time has become shorter compared with conventional mechanics that need multiloop archwires. However,

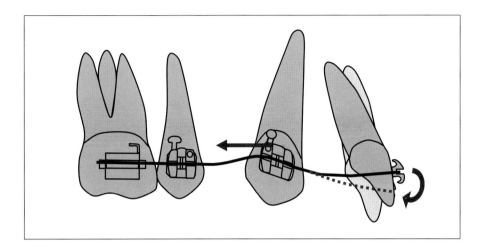

Fig 4-9 Canine distalization along a flexible archwire or excessive force may cause extrusion of the incisors and deepening of the bite owing to excessive distal tipping.

straight wires can be affected by tooth positions, bracket inclinations, and general form of the dental arch, thus offering statically indeterminate mechanics. Furthermore, it may be difficult to predetermine the final inclinations of the occlusal planes. Intermaxillary vertical elastics or headgear are often needed to keep interocclusal relationships under control throughout treatment. Despite these disadvantages, straight wires and frictional mechanics are the most commonly used force systems in daily practice because they are easy to apply and require less chair time.

In frictionless mechanics, the teeth are moved as groups or individually with loops. This enables the clinician to avoid friction between brackets and archwires, which, in space closure, will slow or retard long-range tooth movement. In addition, controlling the amount of forces and moments makes it possible to control adverse mechanical effects that may be unavoidable with straight wires.

Advantages of the frictional system

- Straight wires are easy to apply, thus requiring less chair time.
- Patient discomfort (hygiene problems, soft tissue irritation) is less common compared with looped wires.
- Leveling can be performed easily with highly flexible NiTi wires.
- The whole dental arch can be controlled with only one archwire.

Disadvantages of the frictional system

- Friction is a multifactorial phenomenon that makes mechanics unpredictable. Any interaction between wire, brackets, and ligatures causes friction; therefore, anchorage loss is more likely to occur with frictional mechanics.
- Occlusal plane inclination and interocclusal relationships may need to be controlled by intermaxillary elastics, microimplants, or headgear.
- Canine distalization along flexible archwires or excessive force application may cause extrusion of incisors, resulting in anterior deep bite (Fig 4-9).
- En masse retraction is difficult without headgear, which requires considerable patient cooperation.

Advantages of the frictionless system

- Effectively increases moment-to-force ratios by means of loops. This allows for torque control of anterior teeth during space closure.
- Lengthens the distance between points of force application, thereby reducing the wire's load/deflection rate and increasing its working range.
- Offers more predictable mechanics in which amounts of force and moment are measurable.
- Selective mechanics such as incisor intrusion and molar uprighting are easier to perform.

Disadvantages of the frictionless system

- Loop bends require significant chair time.
- Loops can be uncomfortable for the patient and may cause hygiene problems.
- Transverse control of the canines during distalization is reduced compared with sliding mechanics.

Orthodontic treatment requires controlled movement through force control and patience. Whichever system is chosen, some unpredictable adverse mechanical effects during treatment are to be expected. However, the main goal can be achieved if biomechanical principles are applied correctly.

References

1. Nikolai RJ. Bioengineering Analysis of Orthodontic Mechanics. Philadelphia: Lea & Febiger, 1985.
2. Tanne K, Matsubara S, Hotei Y, Sakuda M, Yoshida M. Frictional forces and surface topography of a new ceramic bracket. Am J Orthod Dentofacial Orthop 1994;106:273–278.
3. Drescher D, Bourauel C, Schumacher HA. Frictional forces between bracket and arch wire. Am J Orthod Dentofacial Orthop 1989;96:397–404.
4. Angolkar PV, Kapila S, Duncanson MG Jr, Nanda RS. Evaluation of friction between ceramic brackets and orthodontic wires of four alloys. Am J Orthod Dentofacial Orthop 1990;98:499–506.
5. Kusy RP, Whitley JQ. Effects of surface roughness on the coefficients of friction in model orthodontic systems. J Biomech 1990;23:913–925.
6. Stannard JG, Gau JM, Hanna M. Comparative friction of orthodontic wires under dry and wet conditions. Am J Orthod 1986;89:485–491.
7. Frank CA, Nikolai RJ. A comparative study of frictional resistances between orthodontic bracket and arch wire. Am J Orthod 1980;78:593–609.
8. Tosun Y, Ünal H. Study of friction between various wire and bracket materials. Turk Ortodonti Derg 1998;11:35–48.
9. Peterson L, Spencer R, Andreasen GF. Comparison of frictional resistance of Nitinol and stainless steel wires in Edgewise brackets. Quintessence Int Digest 1982;13:563–571.
10. Kapila S, Angolkar PV, Duncanson MG Jr, Nanda RS. Evaluation of friction between edgewise stainless steel brackets and orthodontic wires of four alloys. Am J Orthod Dentofacial Orthop 1990;98:117–126.
11. Ogata RH, Nanda RS, Duncanson MG Jr, Sinha PK, Currier GF. Frictional resistances in stainless steel bracket-wire combinations with effects of vertical deflections. Am J Orthod Dentofacial Orthop 1996;109:535–542.
12. Andreasen GF, Quevedo FR. Evaluation of frictional forces in the 0.022 × 0.028 edgewise bracket in vitro. J Biomech 1970;3:151–160.
13. Tidy DC. Frictional forces in fixed appliances. Am J Orthod Dentofacial Orthop 1989;96:249–254.
14. Bednar JR, Gruendeman GW, Sandrik JL. A comparative study of frictional forces between orthodontic brackets and arch wires. Am J Orthod Dentofacial Orthop 1991;100:513–522.
15. Omana HM, Moore RN, Bagby MD. Frictional properties of metal and ceramic brackets. J Clin Orthod 1992;26:425–432.
16. Garner LD, Allai WW, Moore BK. A comparison of frictional forces during simulated canine retraction of a continuous edgewise arch wire. Am J Orthod Dentofacial Orthop 1986;90:199–203.
17. Riley JL, Garrett SG, Moon PC. Frictional forces of ligated plastic and metal edgewise brackets [abstract 21]. J Dent Res 1979;58:98.
18. Edwards GD, Davies EH, Jones SP. The ex vivo effect of ligation technique on the static frictional resistance of stainless steel brackets and archwires. Br J Orthod 1995;22:145–153.
19. Popli K, Pratten D, Germane N, Gunsolley J. Frictional resistance of ceramic and stainless steel orthodontic brackets [abstract 747]. J Dent Res 1989;68:275.
20. Berger JL. The influence of the SPEED bracket's self-ligating design on force levels in tooth movement: A comparative in vitro study. Am J Orthod Dentofacial Orthop 1990;97:219–228.
21. Shivapuja PK, Berger J. A comparative study of conventional ligation and self-ligation bracket systems. Am J Orthod Dentofacial Orthop 1994;106:472–480.
22. Stoner M. Force control in clinical practice. Am J Orthod 1960;46:163–186.
23. Buck TE, Scott JE, Morrison WE. A Study of the Distribution of Force in Cuspid Retraction Utilizing a Coil Spring [thesis]. Houston: University of Texas, 1963.
24. Yamaguchi K, Nanda RS, Morimoto N, Oda Y. A study of force application, amount of retarding force, and bracket width in sliding mechanics. Am J Orthod Dentofacial Orthop 1996;109:50–56.
25. Tosun Y, Ünal H, Türkoğlu K. Effect of angulation between bracket and wire on the frictional resistance. Presented at the International Congress of the Turkish Orthodontic Society, Istanbul, 16–20 June 1998.
26. Tselepis M, Brockhurst P, West VC. The dynamic frictional resistance between orthodontic brackets and arch wires. Am J Orthod Dentofacial Orthop 1994;106:131–138.
27. Dickson JA, Jones SP, Davies EH. A comparison of the frictional characteristics of five initial alignment wires and stainless steel brackets at three bracket to wire angulations—An in vitro study. Br J Orthod 1994;21:15–22.

28. Saunders CR, Kusy RP. Surface topography and frictional characteristics of ceramic brackets. Am J Orthod Dentofacial Orthop 1994;106:76–87.
29. Kusy RP, Whitley JQ, Prewitt MJ. Comparison of the frictional coefficients for selected archwire-bracket slot combinations in the dry and wet states. Angle Orthod 1991;61:293–302.
30. Pratten DH, Popli K, Germane N, Gunsolley JC. Frictional resistance of ceramic and stainless steel orthodontic brackets. Am J Orthod Dentofacial Orthop 1990;98:398–403.
31. Baker KL, Nieberg LG, Weimer AD, Hanna M. Frictional changes in force values caused by saliva substitution. Am J Orthod Dentofacial Orthop 1987;91:316–320.
32. Bazakidou E, Nanda RS, Duncanson MG Jr, Sinha PK. Evaluation of frictional resistance in esthetic brackets. Am J Orthod 1997;112:138–144.
33. Schwartz ML. Ceramic brackets. J Clin Orthod 1988;22:82–88.
34. Kusy RP, Whitley JQ. Coefficients of friction for arch wires in stainless steel and polycrystalline alumina bracket slots. I. The dry state. Am J Orthod Dentofacial Orthop 1990;98:300–312.
35. Tosun Y, Ünal H, Şen BH. Evaluation of the frictional resistance of stainless steel brackets manufactured with different techniques. Presented at the International Congress of the Turkish Orthodontic Society, Istanbul, 16–20 June 1998.
36. Vaughan JL, Duncanson MG Jr, Nanda RS, Currier GF. Relative kinetic frictional forces between sintered stainless steel brackets and orthodontic wires. Am J Orthod 1995;107:20–27.
37. Proffit WR. Contemporary Orthodontics. St Louis: Mosby, 1986: 246,264.
38. Burstone CJ. The biomechanical rationale of orthodontic therapy. In: Melsen B (ed). Current Controversies in Orthodontics. Chicago: Quintessence, 1991:131–146.
39. Burstone CJ, Hanley KJ. Modern Edgewise Mechanics, Segmented Arch Technique. Glendora, CA: Ormco, 1989:5–16.
40. Burstone CJ. Rationale of the segmented arch. Am J Orthod 1962;48:805–822.
41. Burstone CJ. The mechanics of the segmented arch techniques. Angle Orthod 1966;36:99–120.

Anchorage Control

Anchorage is resistance against undesired tooth movement. In modern orthodontics, anchorage loss can be a significant complication during treatment. Therefore, anchorage control is an important issue that needs to be addressed right from the leveling stage.

Uncontrolled tipping is the easiest tooth movement to accomplish with orthodontic appliances, whereas root movement is the most difficult and complicated. The concept of anchorage preparation has been used for several years in the Tweed technique to reinforce anchorage through anchorage bends before anterior retraction. In those cases wherein anchorage bends are not sufficient, the anchorage teeth must be actively reinforced with auxiliaries. This anchorage reinforcement can be achieved in several ways.

Intraoral Methods

Increasing the number of teeth

The simplest and most practical way of reinforcing anchorage is to increase the number of teeth by binding them together with a figure-eight ligature. The number of roots and the root surface areas affect the anchorage capacity of a tooth. Theoretically, a three-rooted tooth has more anchorage value than a single-rooted tooth[1] (Fig 5-1). The type of movement (ie, tipping or root movement), however, defines the real capacity of the anchorage of a tooth. Under the effect of root movement, the anchorage value of a single-rooted tooth (ie, maxillary canine) can overcome that of a maxillary molar (ie, the rowboat effect; see chapter 3). Clinically, as far as anchorage is concerned, one should not fully rely on the number of teeth and roots, but on the force system acting on them.

The Nance appliance

The Nance appliance is an auxiliary arch used in conjunction with fixed appliances in moderate anchorage cases. The Nance appliance has an acrylic button placed anteriorly at the deepest point of the palate. The two ends of the arch are either soldered to the palatal aspects of the maxillary molar bands or are inserted into the palatal tubes. The shape and depth of the palate plays an important role in the retention of the Nance appliance. If the palate is shallow anteriorly, the acrylic button may slide on the palatal mucosa, so narrower and deeper palates are more suitable for this type of retention. The acrylic button should be prepared wide enough to cover the anterior part of the palate. A small button will not provide enough anchorage support and may puncture the palatal mucosa.

5 | Anchorage Control

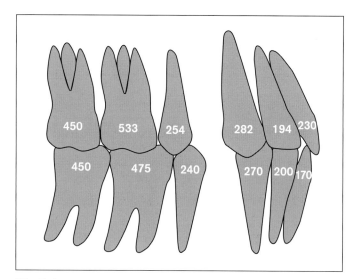

Fig 5-1 Anchorage values of various teeth.[1] (Redrawn from Proffit[1] with permission.)

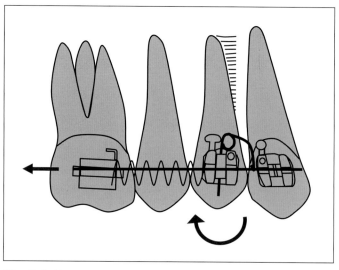

Fig 5-2 Uprighting springs increase the resistance of premolars against mesially directed forces during molar distalization, and a clockwise moment by the uprighting spring creates distal crown movement of the first premolar.

Even though the mucosa in contact with the acrylic button is resistant and covered with a thick keratin layer, it might not resist high, continuous forces. It has been shown that chronic inflammation has developed on the mucosa under the acrylic button due to heat and lack of hygiene.[2] The force, therefore, should be optimal, and the Nance appliance should be removed as soon as it is no longer needed for anchorage.

Uprighting springs

Uprighting springs are used to reinforce the resistance of the first premolars, which serve as anchorage in molar distalization with a combination of superelastic nickel-titanium (NiTi) coil springs and the Nance appliance. When these springs are activated after being placed in the vertical slots of the premolar brackets, they apply mesial root torque, creating resistance against mesially directed forces by the coil springs (Fig 5-2).

Sliding jig

A sliding jig, used in combination with intramaxillary and intermaxillary elastics, can be accepted as an anchorage reinforcement auxiliary. Consisting of a 0.7-mm wire with a hook, the sliding jig slides along the main archwire and transmits the Class II elastic force directly to the molar. The mechanics of this appliance are explained in detail in chapter 8.

Cortical bone anchorage

The principle of the cortical bone method is based on the biologic differences between cortical and trabecular bone. Teeth move more easily in trabecular bone than in cortical bone. If the roots of the anchorage teeth are in cortical bone, their resistance to movement is increased. Ricketts[3] has suggested putting active lingual root torque on mandibular canines to place them in trabecular bone before distalization to make movement smoother.

Active buccal root torque can be placed in molars to increase their anchorage against mesially directed forces. A transpalatal arch can be used for this purpose. Both ends of the arch should be inserted at equal angles into the molar tubes to obtain equal and opposite moments and avoid vertical balancing forces (Fig 5-3).

Transpalatal arch

Transpalatal arches reinforce posterior anchorage by incorporating maxillary right and left molars (see chapter 8). Root[4] stated that in a case involving a transpalatal arch, the anchorage value needed to correct a maloc-

Intraoral Methods

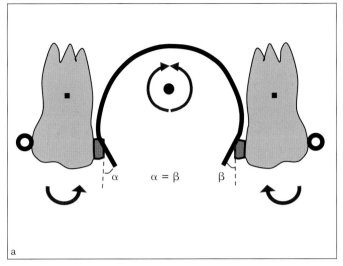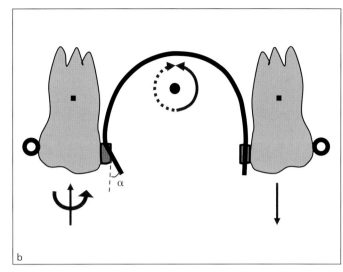

Fig 5-3 Cortical bone anchorage prevents the molars from moving mesially. A transpalatal arch moves the roots by reciprocal anchorage. *(a)* Obtaining equal and opposite moments is important to avoid vertical balancing forces. *(b)* Unequal moments may cause an imbalance in the occlusion.

Fig 5-4 The lip bumper reinforces anchorage of the mandibular molars using mentalis muscle force.

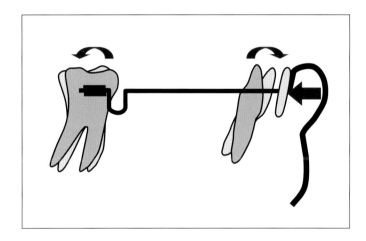

clusion in the mandible decreases by 1 mm in approximately 1 year. Soysal[5] claims that the transpalatal arch can be an alternative to headgear even in cases needing strong anchorage. However, the transpalatal arch cannot resist anteroposterior forces as effectively as headgear, but it can be used in moderate anchorage cases.

Lip bumper

The lip bumper is used to reinforce anchorage of mandibular molars by taking support from mentalis muscle activity. In fact, it is a myofunctional appliance used to eliminate the compressing effect of the mentalis muscle forces on the mandibular dental arch, especially in the transitional dentition. Retraction of the mentalis and buccinator muscles causes an expansion of the mandibular arch at the premolar and molar area and protrusion of the incisors by the tongue forces. Expansion of the molars can also be obtained by widening the arms of the arch. Because the force is applied directly onto the mandibular molars, it helps reinforce molar anchorage and even helps upright them (Fig 5-4). As the anteroposterior and transverse sizes of the mandibular arch are increased, mild crowding in the incisor area can be corrected spontaneously.[6,7] Size and stability of arch form obtained with the lip bumper can be maintained with a lingual arch until eruption of the permanent dentition.

5 | Anchorage Control

Extraoral Appliances

Extraoral appliances have been the strongest and most reliable anchorage reinforcement method for several years. Approximately 300 to 350 g is sufficient force to prevent the posterior teeth from moving mesially.[8] Patient cooperation is key to the success of headgear treatment, and headgear timers can be helpful in that area.[9–11]

Extraoral appliances can be categorized as those that apply forces from anterior to posterior in Class II cases and those that apply forces from posterior to anterior in Class III cases. This chapter focuses on Class II types that apply force by means of a facebow, which is used in most cases.

The objectives of using extraoral appliances are listed under five headings:

1. Anchorage reinforcement
2. Creating an orthopedic effect on the jaws
3. Providing vertical control of the face
4. Changing the occlusal plane inclination
5. Tooth movement

Anteroposterior force application

In skeletal Class II cases, the main purpose of extraoral force is to control or redirect forward and downward growth of the maxilla. Besides its skeletal effects, extraoral force also has dental effects such as distalizing maxillary molars to gain space on the dental arch and reinforce the anchorage. Extraoral appliances can be used either on their own or combined with fixed appliances, usually on the maxillary first molars.[12–14]

Extraoral appliances are classified by different authors according to their direction of force, length, or angulation of their outer arms. Facebows of various dimensions are manufactured according to dental arch sizes. These are used with short, medium, or long outer arms, depending on the needs of each patient. The aim here is not to prescribe, but rather to state the general mechanical principles of extraoral forces, helping orthodontists select suitable appliances for their particular patients.

The main criterion in selecting extraoral appliances is the vertical growth pattern of the face. In low-angle cases, cervical-pull forces (cervical headgear) should be used because the direction of pull is backward and downward. The vertical component of this force vector pulls the molars down, causing clockwise rotation of the mandible. This leads to bite opening,[8] which is desirable in low-angle patients. In patients with high condylar growth potential, the effect of cervical headgear can be compensated for by condylar growth.[15]

In high-angle cases, a high-pull or vertical-pull extraoral force, which applies an upward pull, must be used. The pulling direction of these appliances is backward and upward. This force direction provides vertical control of the posterior segment by preventing eruption of the molars and sometimes intruding them. In high-angle cases, this prevents clockwise rotation of the mandible and even promotes counterclockwise rotation. For this to occur, however, the patient must have sufficient condylar growth potential.

Analysis in the sagittal plane

Extraoral forces are applied between either the neck or different parts of the head and the outer arms of the facebow and transferred to the molars through the inner arms. The relationship between the line of action of the force and the center of resistance of the molars determines the movements of these teeth.

Cervical headgear

The application of cervical headgear on the molar tubes is shown in Fig 5-5. The direction of force is backward and downward; therefore, the force has horizontal and vertical components. If translation of the molars is indicated, the line of action of the force should pass through the center of resistance of the molars (Fig 5-5a). If the line of action passes below the center of resistance, the molar undergoes clockwise rotation as well as translation (Fig 5-5b). This will cause the crown to move distally while the roots move mesially. If the line of action of the force passes above the center of resistance (Fig 5-5c), the molar rotates counterclockwise as well as translates.

High-pull headgear

High-pull force passing through the center of resistance causes translation with intrusion (Fig 5-6a), while force passing below the center of resistance causes clockwise

Extraoral Appliances

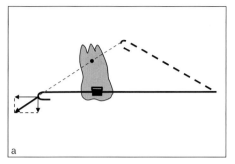 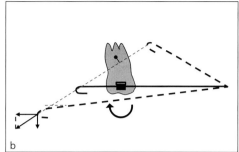 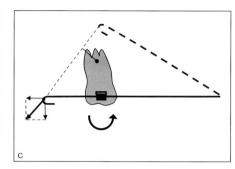

Fig 5-5 The direction of force in cervical headgear is backward and downward. *(a)* If the line of action of the force passes through the center of resistance of the molar, translation occurs in the direction of the force. *(b)* If the force passes below the center of resistance of the molar, a clockwise moment occurs. *(c)* If it passes above the center of resistance, a counterclockwise moment occurs. Note that the molar movement is determined by its relation with the center of resistance rather than the length of the outer arms or their angulations. As seen in Fig 5-5b, the results obtained with the short arm angled upward, the medium-length arm nonangulated, or the long arm angled downward are mechanically the same because they are on the same line of action.

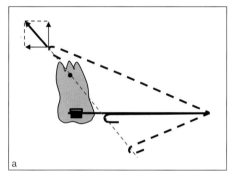 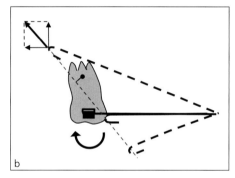 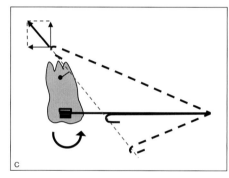

Fig 5-6 Direction of force in a high-pull headgear is backward and upward. *(a)* Force passing through the center of resistance causes translation. *(b)* Force passing below the center of resistance causes clockwise rotation. *(c)* Force passing above the center of resistance causes counterclockwise rotation.

rotation (Fig 5-6b), and force passing above the center of resistance causes counterclockwise rotation of the molars (Fig 5-6c).

Adjustment to the direction of force

Because the center of resistance of the molar is a stable point, to change the direction of the resultant force of an extraoral appliance, it is necessary to change either the point of support (the neck in cervical headgear and the parietal bone in high-pull headgear) or the length or angle of the facebow. It is easier to make this change on the head than on the neck. It is nearly impossible to change the point of application of cervical headgear, but on high-pull headgear, the point of support can be moved slightly upward or downward. For example, in skeletal open bite cases, if molar intrusion is desired, the vertical vector should be as high as possible (Fig 5-7), moving the point of support higher on the head. In this case, because the vertical component of force is much higher than the horizontal, the tooth will move upward rather than backward.

If distal translation of the molars is indicated, the horizontal vector should be increased. To achieve this, the line of action of the force must pass between the parietal bone and the neck (the support points of high-pull and cervical headgears). When applying these types of extraoral appliances (called *straight-pull* or *combined headgear*),

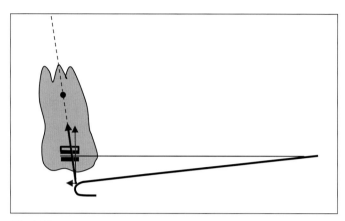

Fig 5-7 In cases requiring vertical control or molar intrusion, it is necessary to increase the vertical component of force applied by high-pull headgear. To achieve this, the point of support should be moved higher on the head.

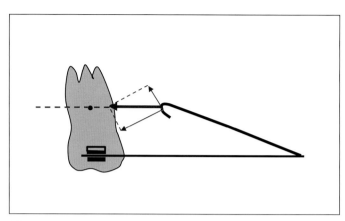

Fig 5-8 A straight-pull headgear, which is a combination of high-pull and cervical headgears, can be used when distal translation of the molar is desired.

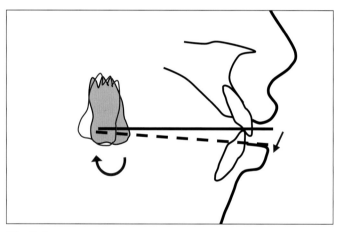

Fig 5-9 When the molar tips distally, the front of the facebow moves downward and presses upon the lower lip.

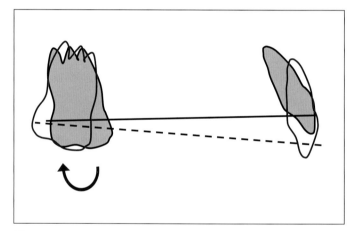

Fig 5-10 Excessive distal tipping of the molars can cause extrusion of the incisors and change the inclination of the occlusal plane.

the resultant of the upward and downward force vectors must pass through the center of resistance of the molar (Fig 5-8).

In practice, translation of molars is a fairly difficult movement because of the difficulty in determining a tooth's center of resistance, which is theoretically accepted to be at the trifurcation of a three-rooted tooth. Two practical methods can be suggested to determine this point. The first one is to mark the position of the molar on the cheek simply by visual determination. The second is to mark it on the cephalometric film taken by applying a thin ligature wire between the outer arm and the point of application (ie, the neck for cervical headgear and the occipital bone for high-pull headgear) while the facebow is in the mouth. Intraoral adjustment of the facebow can be done by measuring the distance between the line of action (the ligature wire) and the center of resistance on the cephalometric film.

Even if the center of resistance can be accurately determined using these methods, the relationship between the line of action of the force and the center of resistance will change (and so will the force system) shortly

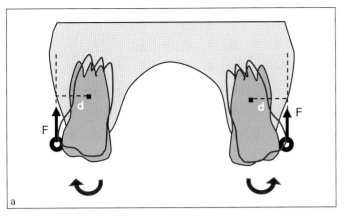 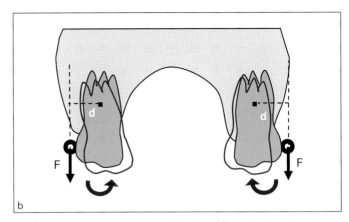

Fig 5-11 *(a)* High-pull headgear force (F) applied upward on the buccal aspect of the molars can cause buccal tipping. *(b)* Cervical headgear force applied downward on the buccal aspect of the molars can cause palatal tipping and posterior crossbite. d, distance.

after the tooth starts to move. In practice, checking the movements of the molars on a regular basis is essential to keep the force and movement under control.

Practically, the line of action of the force can be adjusted by bending the outer arms of the facebow upward or downward, depending on the particular needs of the patient. The vertical aspect of the front of the facebow indicates the type of movement of the molar. A downward movement of the facebow against the lower lip indicates distal crown tipping (Fig 5-9). This also indicates that the line of action of the force passes below the center of resistance of the molar. To correct this, bend the outer arms slightly upward. To check whether this bend is correct, observe the movement of the front of the facebow when force is applied. With correct angulation, the facebow will show a slightly upward deflection after force is applied. If this deflection is excessive, the angulation is too high. On the other hand, if the molar is tipped mesially, the facebow will move upward and touch the upper lip. To correct this, bend the outer arms downward for the line of action of the force to pass below the center of resistance. When force is applied to the bow, it must deflect slightly downward. Force and movement control should be addressed regularly to avoid excessive molar tipping, because it can affect the inclination of the maxillary occlusal plane (Fig 5-10).

Analysis in the frontal plane

Extraoral forces are applied to the molars through the extraoral tubes located on the buccal aspects of these teeth. Because these tubes are attached at a certain distance from the centers of resistance of the molars, a moment will occur during vertical movements (ie, extrusion or intrusion). When high-pull headgear is used, buccal expansion of the molars due to this moment is a common side effect. Sometimes this expansion can even cause a buccal nonocclusion[16] (Fig 5-11a). During buccal tipping of maxillary molars, the palatal cusps move down and cause premature contacts with their mandibular antagonists. In high-angle cases, premature contacts may cause clockwise rotation of the mandible, combined with an anterior open bite and a convex profile. In such cases, use of a transpalatal arch to provide buccolingual control of the molars is recommended. A transpalatal arch can also help control molar vertical movements by transferring the vertical pressure applied by the tongue during swallowing (see chapter 8).

On the other hand, a downward force vector from cervical headgear causes the molars to extrude and tip palatally, which may lead to posterior crossbite (Fig 5-11b). To eliminate this adverse effect, the inner arms of the facebow should be widened slightly before inserting into the tubes; it is even more effective to use a transpalatal arch.

5 | Anchorage Control

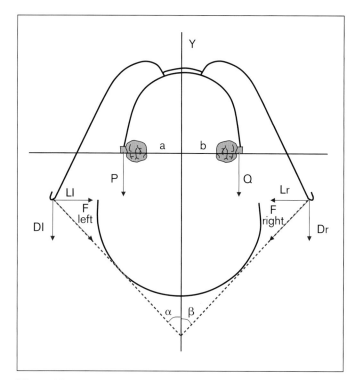

Fig 5-12 In a symmetric headgear appliance, the magnitudes of the right (F right) and left (F left) forces are equal. Because the lateral components of the resultant forces (Ll and Lr) are equal, they balance each other out. Therefore, only the distal forces (P and Q) are effective on the molars. Dr and Dl, right and left resultant distal force vectors; Y, central line of the facebow; a and b, two equal sections of the intraoral arch; α and β, angles that occur between Y and F left and F right.

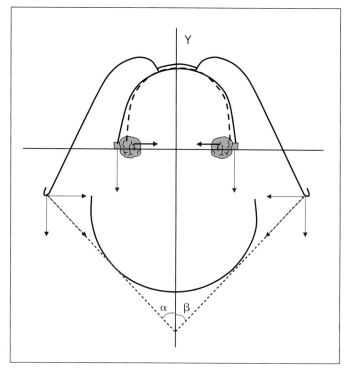

Fig 5-13 If the joint between the inner arch and outer arms of the facebow is not rigid enough, lateral force components may deflect the inner arch, causing constriction of the maxillary dental arch and posterior crossbite.

Analysis in the transverse plane

Widening of the inner bow is also required to prevent crossbite as the molars move into a wider portion of the arch. With the high-pull headgear, however, it is not necessary to widen the inner arms because transverse expansion will occur as a result of the intrusive force. Figure 5-12 shows a symmetric extraoral appliance that applies forces with equal magnitudes on the outer arms of the facebow. In such a facebow, whose outer arms are assumed to be firmly soldered to the inner bow, the right and left resultant forces (F right and F left) intersect at the central line (Y) of the facebow, and the intraoral arch is separated into two equal sections (a and b). Because the lengths of the outer arms are equal, the α and β angles that occur between F right and F left and Y are also the same. The lateral components balance each other because they are equal and opposite; as a result, there are no lateral forces on the inner arms of the facebow, and only the distally directed Q and P forces applied to the molars remain in the system.

In a facebow in which the junction of the outer arms and the inner arch is not rigid, as the outer arms deflect with the effect of the force, the inner arch contracts and causes the dental arch to narrow (Fig 5-13). The junction between the outer arms and inner arch, therefore, must be rigid enough to avoid this undesired side effect.

Application of extraoral forces to the entire dental arch

So far, the extraoral forces discussed in this chapter have been applied only to the maxillary first molars. Because a fixed appliance consists of multiple brackets, a picture different from the foregoing is expected.

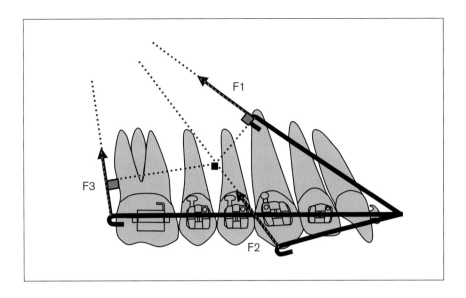

Fig 5-14 The center of resistance in the maxilla is located between the roots of the premolars. The relationship of this center to the extraoral force determines the type of movement of the whole arch. If backward and upward movement is indicated, the line of action of the extraoral force should pass through the center of resistance (F2). A force passing above the center of resistance (F1) causes the whole dental arch to rotate counterclockwise. If the force passes below the center of resistance (F3), it will cause clockwise rotation.

In a maxillary dental arch incorporated with a stiff rectangular archwire, the center of resistance is located between the roots of the first and second premolars (Fig 5-14). Extraoral force is again applied to the maxillary molars. The relationship of the extraoral force to the center of resistance of the maxillary "block" determines the type of movement of the whole dental arch. If the force passes though the center of resistance of the dental arch (F2), it will translate backward and upward. This is the desired type of movement for most skeletal Class II patients with a normal vertical pattern.

If the line of action of the force passes above the center of resistance (F1), the dental arch will be under the influence of a counterclockwise moment, which results in extrusion of the molars and intrusion of the incisors. Molar extrusion will cause the bite to open with clockwise rotation of the mandible. This approach is indicated in skeletal Class II, low-angle cases with anterior deep bite. Force passing below the center of resistance will cause the dental arch to rotate clockwise, resulting in molar intrusion and incisor extrusion (F3). This is the preferred approach for high-angle cases with anterior open bite.

Application of asymmetric forces

Asymmetric extraoral force is needed in Class II subdivision cases to distalize the Class II side and obtain a Class I relationship. There are various types of extraoral appliances that apply asymmetric force; only the most effective and practical types are discussed here.

Extraoral appliances with asymmetric arms

In extraoral appliances in which one arm is longer than the other (Fig 5-15), the α and β angles, formed by the equal F left and F right resultant forces with the central line Y, differ. Because the β angle is mathematically higher than the α angle, the magnitude of the Q force acting on the right molar will be higher than that acting on the left molar. Furthermore, there are differences between the lateral components Lr and Ll of the F right and F left resultant forces. Thus, a lateral force vector to the left will occur on the right molar as a result of this difference.[17]

When it is necessary to apply an asymmetric extraoral force, the outer arm on the side where more force is needed must be kept longer than the other outer arm.

5 | Anchorage Control

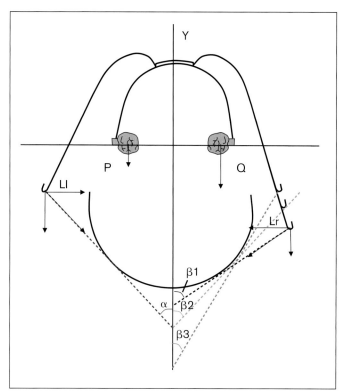

Fig 5-15 In asymmetric headgear appliances, one of the outer arms is longer than the other. The magnitude of the distal force (Q) effective on the molar by the long arm is higher than that applied by the short arm (P). Because the lateral component (Lr) of the resultant force applied to the side of the long arm is greater than the lateral component (Ll) of the short arm, there is a strong palatal force in effect on the longer-arm side, carrying the risk of forcing the molar on that side into crossbite while it is being moved distally. Y, central line of the facebow; α and β, angles that occur between Y and the left and right resultant forces. (Modified from Oosthuizen et al[17] with permission.)

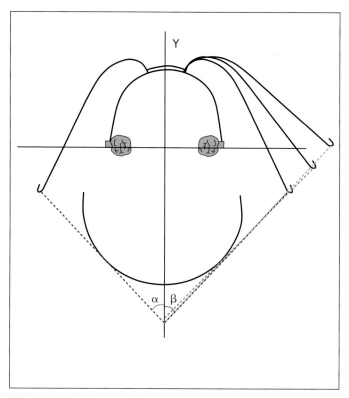

Fig 5-16 Bending one of the outer arms laterally has only a small effect on the asymmetric force vector. Y, central line of the facebow; α and β, angles that occur between Y and the left and right resultant forces. (Modified from Oosthuizen et al[17] with permission.)

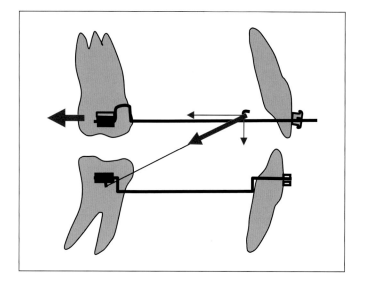

Fig 5-17 The force of Class II elastics from the canine hooks of the facebow both increases the effect of the extraoral force and eliminates the extrusive effect of conventional Class II elastics on the maxillary anterior teeth.

Fig 5-18 With reverse headgear, because the force is applied to the hooks (usually located at the level of the lower lip), the magnitude of the force applied to the chin is higher than that applied to the forehead.

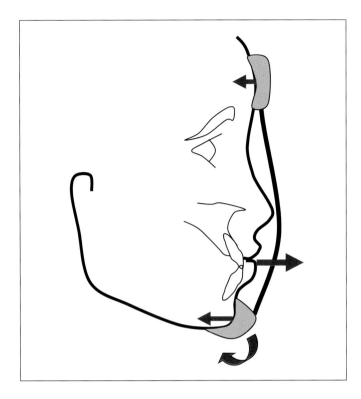

Even though it might be helpful to expand the longer outer arm laterally, this will have minimal effect (Fig 5-16). In asymmetric force application, there is a constriction tendency on the molar side that receives the more distal force. Thus, this molar could shift into crossbite with the effect of this lateral force. If one tries to keep the inner arch wider to avoid this adverse effect, the other molar may expand into buccal nonocclusion. In such applications, excessive forces must be avoided and the use of a transpalatal arch is recommended for better molar control.[17,18]

Canine hooks

Hooks soldered on the facebow at the level of the canines (called *canine hooks*) are used to accept Class II elastics to increase the distalization effect of the extraoral force on the maxillary molar and to help avoid the extrusive effect of conventional Class II elastics placed on the maxillary anterior teeth (Fig 5-17).

Reverse headgears

Reverse headgears are also known as *protraction headgears* or *orthopedic facemasks*. They are principally used in the following situations:

- Skeletal Class III cases with maxillary retrusion or micrognathia
- Minimum anchorage cases

The idea of protruding the maxilla (introduced by Oppenheim in 1944) was effectively brought to life with an orthopedic facemask developed by Delaire at the end of the 1960s. This appliance was then used in different forms by various researchers without any alteration in its fundamental mechanism.[19,20]

Delaire's orthopedic facemask, which is at present the most commonly used reverse headgear, consists of three main sections (Fig 5-18):

- A section taking support from the forehead
- A section taking support from the tip of the chin
- A prelabial arch to which force is applied

Approximately 800 to 1,500 g of force can be applied to achieve an orthopedic effect from reverse headgears. To apply forces of such high magnitude, the maxillary teeth must be tied together tightly by means of rectangular wire or cast in a maxillary splint.[21]

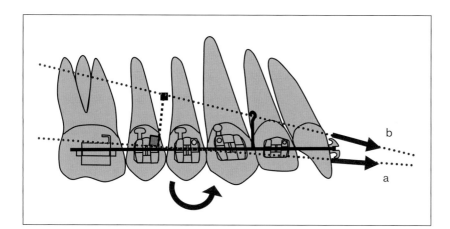

Fig 5-19 In reverse headgear applications, if the line of action of the force passes below the center of resistance of the upper dental arch (a), it causes counterclockwise rotation of the whole arch. This results in extrusion of the molars and an anterior open bite. To avoid this adverse effect, the force must be applied to the power hooks between the canines and lateral incisors so that the line of action passes through the center of resistance or above it (b).

Most patients with maxillary retrusion or micrognathia exhibit serious maxillary constriction. In these cases, it is necessary to expand the maxilla using rapid maxillary (or palatal) expansion before protraction with reverse headgear. With rapid maxillary expansion, point A is expected to come slightly forward.[22,23] The orthopedic forces of reverse headgear can be applied on either the hooks of a cast rapid-expansion appliance or on hooks crimped (or soldered) on a stiff rectangular archwire.

With reverse headgear, the maxilla moves forward by rotating counterclockwise around a point located near the frontomaxillary suture. During this process, the maxillary dental arch advances on the basal bone in a manner called *drawer movement*.[24-27] Even though the whole dental arch is tied with a stiff rectangular archwire and undergoes an orthopedic effect, it is impossible to avoid some dental changes. Protraction of the maxillary incisors is particularly common.

During protrusion of the maxilla, certain changes also occur in the mandible. Owing to the distal force applied to the tip of the chin, the mandible rotates clockwise,[24-27] leading to overbite reduction and a convex soft tissue profile. This effect is not desired in high-angle cases. Therefore, in reverse headgear applications, the vertical facial pattern of the patient must be taken into consideration. In particular, applications in high-angle cases should be avoided.

When reverse headgear is used in cases with a high-angle tendency, the point of application and the direction of the force are even more important. When the force is applied to the molars, the line of action passes below the center of resistance of the maxillary dental arch and causes counterclockwise rotation of the whole arch, resulting in extrusion of the molars and opening of the bite (*a* in Fig 5-19). To avoid this, the line of action of the force should pass through or slightly above the center of resistance of the maxillary arch by way of power hooks placed between the canines and lateral incisors and angulated approximately 30 to 45 degrees to the maxillary occlusal plane (*b* in Fig 5-19).

As the maxillary anterior teeth move forward in reverse headgear application, the mandibular incisors tend to move lingually with the chin cap pressure.[28] To prevent possible mandibular incisor crowding, some lingual root torque can be applied to these teeth. Conversely, labial root torque can be given to the maxillary incisors if the clinician does not want to increase their axial inclinations. Fenn[29] and Subtelny[30] have suggested that this torque increases the thickness of the cortical layer and the development of point A and also prevents undesired extrusion of the maxillary incisors.

In minimum anchorage cases, reverse headgear can be applied to the mandibular premolars or molars. In these cases, the line of action of the force should be forward and slightly upward to avoid irritation of the lower lip.

With headgear application, an accurate treatment plan, timing, and patient cooperation are all vital to success. An orthopedic effect can be best achieved before closure of the maxillary sutures and completion of the pubertal growth period. Thus, the results achieved after this period (in young adulthood or adulthood) will be mostly dental.[21,31-33] Because tongue position can cause relapse after reverse headgear application, a functional tongue raiser, in the form of an acrylic plate, can direct the tongue forward and upward and help ensure permanent results.[34]

Temporary Anchorage Devices

Because anchorage plays such an important role in orthodontic mechanics, a reliable source of anchorage for efficient tooth movement has long been sought. Although headgear has been the most reliable and powerful source of anchorage, this device is dependent on patient compliance; should that cooperation be lacking, the success of treatment would be at risk. Temporary anchorage devices (TADs), also called *microimplants*, recently introduced to the world of orthodontics, seem to be reliable intraoral sources of anchorage; thus, they have become a common treatment tool in many clinics worldwide. A number of studies suggest that TADs produce sufficient anchorage against orthodontic forces.[35–41] Their stability and clinical efficiency, however, have been the topics of several additional studies. In this section, the principles of microimplant application and their biomechanical considerations are examined.

Rationale for use of microimplants

There are a number of reasons microimplants are being widely accepted by clinicians, including:

- TADs provide absolute anchorage similar to an ankylosed tooth; thus, undesirable movement in reaction to applied force can be avoided.[42]
- They do not rely on patient cooperation.
- Because of their small size, they can easily be inserted anywhere in the mouth where there is adequate bone support. The insertion requires only a simple surgical process, which can be performed under clinical conditions.
- They can easily be removed when anchorage is no longer needed.
- The fact that microimplants can be inserted anywhere in the mouth offers a variety of mechanical combinations. For example, en masse retraction in a maximum anchorage case, which is extremely challenging with conventional mechanics, could be easily performed using microimplant anchorage mechanics.
- Unlike dental implants, TADs can have orthodontic force applied shortly after insertion, which saves time at the clinic.
- Microimplant anchorage mechanics help correct several challenging problems, such as open bite, deep bite, asymmetry, molar uprighting, and scissors bite. It enables clinicians to focus directly on the main problem without spending time on leveling and alignment. For example, in a high-angle case wherein the mandibular molars have tipped mesially because of loss of premolars, molar uprighting needs to be accomplished by intrusion—a fairly complicated procedure requiring high anchorage and heavy archwires in the anterior segment. Leveling before uprighting usually comprises a lengthy portion of the overall treatment time. Using microimplant anchorage mechanics, however, molar uprighting can be performed independent of other parts of the dental arch without the need for anchorage preparation.

Stability of microimplants

One of the most important requirements of microimplant use is stability during treatment. Treatment success with TADs has been increasing dramatically thanks to cutting-edge technology and improved materials, design, and insertion techniques. Among the factors that affect microimplant stability, thickness and volume of the cortical bone are the most important. Likewise, thickness and length of the implant, amount of force applied, inflammation of peri-implant tissues, and mandibular plane angle need to be taken into consideration during application.[41,43] The length of time after microimplant insertion was not found to be associated with stability, but several studies suggest that the application of force can be a factor.[41–45] Sufficient mechanical interdigitation between cortical bone and the TAD is an important component that affects stability. According to Tsunori et al[46] and Masumoto et al,[47] patients with high mandibular plane angles are at higher risk for instability of microimplants because they have thinner cortical bone in the mandibular first molar region than patients with low mandibular plane angles. They suggest the use of microimplants with a diameter larger than 2.3 mm or miniplates in high-angle cases.

Biomechanical considerations

The same biomechanical principles apply for microimplant anchorage as for conventional mechanics. As mentioned earlier, the fact that TADs can be inserted anywhere in the mouth—there is no anatomical or periodontal restriction—simplifies specific mechanics, which enables orthodontists to perform more efficient treatment in a shorter time compared with conventional techniques.

In microimplant anchorage mechanics, the relationship between the line of action of the force and the center of resistance defines the type of tooth movement. Forces passing through the center of resistance cause translation, and those passing off center cause rotation or tipping (see chapter 1). Therefore, the location of the implant can be determined according to the desired mechanics.

Limitations of microimplant anchorage

Despite its advantages over conventional techniques, microimplant anchorage has certain limitations that need to be considered during treatment planning. Absolute anchorage does not resolve all orthodontic problems. In all cases, what is essential is controlling the amount and direction of force.

Microimplant application has surgical and orthodontic components. Though it requires only minor surgery, important anatomical structures such as the sinuses, palatine neurovascular bundle, mental foramen, and mandibular and lingual nerves; anatomical variations; and distance between roots limit the locations for microimplant insertion. Because roots are conical, the distance between them increases toward the apex. In addition, the alveolar crest is not a good location for TAD insertion. Apically, attached gingiva is a suitable location for patient comfort.

Microimplant biomechanics is also limited by anatomy. Because the TAD is inserted between the roots, somewhere close to the center of resistance of the teeth or dental arch, the direction of force is usually intrusive rather than extrusive. This is an advantage in many cases, but it needs to be monitored throughout treatment because it may cause asymmetry, distort the dental arch, or change the cant of the occlusal plane; it is contraindicated when extrusion is necessary. Nor is transverse canting of the occlusal plane desirable, as it is difficult to correct with conventional mechanics.[42] The location of the microimplant might limit tooth movement, as in molar distalization. If the second premolar root contacts the TAD, the latter should be reinserted close to the molar roots.

The TAD is usually well tolerated by the patient, but it may be uncomfortable in the mandibular canine and incisor area. Likewise, inflammation of the soft tissues because of irritation is common if the TAD is inserted beyond the attached gingiva.

Microimplant anchorage makes long-range tooth movements such as full maxillary arch distalization possible. However, individual or en masse tooth movements are limited by anatomical and biologic structures. Any attempt to push the boundaries of the bone can cause dehiscence or root resorption.

References

1. Proffit WR. Contemporary Orthodontics. St Louis: Mosby, 1986:261,368–377.
2. Tosun Y. Histological study of the effect of Nance appliance on the palatal mucosa. Turk Ortodonti Derg 1992;5:37–40.
3. Ricketts RM. Bioprogressive Therapy. Denver: Rocky Mountain Orthodontics, 1979:100.
4. Root TL. The level anchorage system. In: Graber TM, Swain BF (eds). Orthodontics: Current Principles and Techniques. St Louis: Mosby, 1985:641–663.
5. Soysal M. Comparative Study of the Anchorage Mechanisms with Maxillary Extraction Cases Treated with Straight Wire [thesis]. İzmir, Turkey: Aegean Univ, 1994.
6. Werner SP, Shivapuja PK, Harris EF. Skeletodental changes in the adolescent accruing from use of the lip bumper. Angle Orthod 1994;64:13–20.
7. Atagün C. Application and Effects of Lip Bumper in Class II/1 Cases [thesis]. İzmir, Turkey: Aegean Univ, 1991.
8. Tosun Y. Comparative Study of the Effects of Cervical and High Pull Headgears on Dento-Cranio-Facial Structures of Class II/1 Cases in the Transitional Dentition [thesis]. İzmir, Turkey: Aegean Univ, 1989.
9. Güray E, Orhan M. Selçuk type headgear-timer (STHT)–Introduction and testing of its reliability. Part I: Laboratory work. Turk Ortodonti Derg 1994;7:242–247.
10. Güray E, Orhan M. Selçuk type headgear-timer (STHT)–Clinical application and its effect on cooperation. Part II: Clinical investigation. Turk Ortodonti Derg 1994;7:248–253.

11. Güray E, Orhan M. Selçuk type headgear-timer (STHT). Am J Orthod Dentofacial Orthop 1997;111:87–92.
12. Ulgen M. Cervical headgear and its mechanism of effect. Istanbul Univ Dishekim Fak Derg 1977;11:123–129.
13. Tosun Y, Işiksal E. Study of the effects of high pull headgear application on dento-cranio-facial structures of Class II/1 cases in the transitional dentition. Turk Ortodonti Derg 1991;4:50–54.
14. Siatowski R. The role of headgear in Class II dental and skeletal corrections. In: Nanda R (ed). Biomechanics in Clinical Orthodontics. Philadelphia: Saunders, 1997:131.
15. Schudy FF. Vertical growth versus antero-posterior growth as related to function and treatment. Angle Orthod 1964;35:75–93.
16. Gülyurt M. Extraoral Appliances (Headgear and Chin Cap). Erzurum, Turkey: Atatürk University Faculty of Dentistry, Department of Orthodontics, 1984.
17. Oosthuizen L, Dijkman JF, Evans WGA. A mechanical appraisal of the Kloehn extraoral assembly. Angle Orthod 1973;43:221–232.
18. Jacobson A. A key to understanding extraoral forces. Am J Orthod 1979;75:361–386.
19. Dogan S, Ertürk N. Use of the face mask in the treatment of maxillary retrusion—A case report. Br J Orthod 1991;18:333–338.
20. Battagel JM, Orton HS. A comparative study of the effects of customized facemask therapy or headgear to the lower arch on the developing Class III face. Eur J Orthod 1995;17:467–482.
21. Dogan S. Study of the Effects of the Application of Extraoral Forces in Skeletal Class III Cases on Craniofacial Complex [thesis]. İzmir, Turkey: Aegean Univ, 1987.
22. Davis WM, Kronman JH. Anatomical changes induced by splitting of the midpalatal suture. Angle Orthod 1969;39:126–132.
23. Haas AJ. Long-term posttreatment evaluation of rapid palatal expansion. Angle Orthod 1980;50:189–217.
24. Delaire J. La croissance maxillaire: déductions thérapeutiques. Trans Eur Orthod Soc 1971;3:1–22,81–102.
25. Delaire J, Verdon P, Flour J. Möglichkeiten und Grenzen extraoraler Kräfte in postero-anteriorer Richtung unter Verwendung der orthopädischen Maske. Fortschr Kieferorthop 1978;39:27–45.
26. Delaire J, Verdon P, Kénési MC. Extraorale Zugkräfte mit Stirn-Kinn-Abstützung zur Behandlung der Oberkieferdeformierungen als Folge von Lippen-Kiefer-Gaumenspalten. Fortschr Kieferorthop 1973;34:225–237.
27. Delaire J, Verdon P, Lumineau JP, Cherga-Négréa A, Talmant J, Boisson M. Some results of extra-oral tractions with front-chin rest in the orthodontic treatment of class 3 maxillomandibular malformations and of bony sequelae of cleft lip and palate [in French]. Rev Stomatol Chir Maxillofac 1972;73:633–642.
28. Campbell PM. The dilemma of Class III treatment. Early or late? Angle Orthod 1983;53:175–191.
29. Fenn C. The Clinical and Cephalometric Results of Face Mask Therapy in the Dentofacial Region [thesis]. Rochester, NY: Eastman Dental Center, 1979:1–37.
30. Subtelny JD. Oral respiration: Facial maldevelopment and corrective dentofacial orthopaedics. Angle Orthod 1980;50:147–164.
31. Cozzani G. Extraoral traction and Class III treatment. Am J Orthod 1981;80:638–650.
32. Haskell BS, Farman AG. Exploitation of the residual premaxillary-maxillary suture site in maxillary protraction—An hypothesis. Angle Orthod 1985;55:109–119.
33. Dogan S, Ertürk N. Late stage evaluation of the skeletal Class III cases treated with orthopaedic face mask. Turk Ortodonti Derg 1990;3:134–143.
34. Öztürk Y, Kılıcoglu H. Application of functional tongue raiser in Class III cases treated with Delaire mask. Turk Ortodonti Derg 1990;3:119–124.
35. Sherwood KH, Burch JG, Thompson WJ. Closing anterior open bites by intruding molars with titanium miniplate anchorage. Am J Orthod Dentofacial Orthop 2002;122:593–600.
36. Park HS, Bae SM, Kyung HM, Sung JH. Micro-implant anchorage for treatment of skeletal Class I bialveolar protrusion. J Clin Orthod 2001;35:417–422.
37. Bae SM, Park HS, Kyung HM, Kwon OW, Sung JH. Clinical application of micro-implant anchorage. J Clin Orthod 2002;36:298–302.
38. Yao CC, Wu CB, Wu HY, Kok SH, Chang HF, Chen YJ. Intrusion of the overerupted upper left first and second molars by mini-implants with partial-fixed orthodontic appliances: A case report. Angle Orthod 2004;74:501–507.
39. Umemori M, Sugawara J, Mitani H, Nagasaka H, Kawamura H. Skeletal anchorage system for open bite correction. Am J Orthod Dentofacial Orthop 1999;115:166–174.
40. Costa A, Raffainl M, Melsen B. Miniscrews as orthodontic anchorage: A preliminary report. Int J Adult Orthodon Orthognath Surg 1998;13:201–209.
41. Miyawaki S, Koyama I, Inoue M, Mishima K, Sugahara T, Takano-Yamamoto T. Factors associated with the stability of titanium screws placed in the posterior region for orthodontic anchorage. Am J Orthod Dentofacial Orthop 2003;124:373–378.
42. Lee JS, Kim JK, Park YC, Vanarsdall RL Jr. Applications of Orthodontic Mini-Implants. Chicago: Quintessence, 2007:111,114.
43. Sung JH, Kyung HM, Bae SM, Park HS, Kwon OW, McNamara JA Jr. Microimplants in Orthodontics. Daegu, Korea: Dentos, 2006:15–32.
44. Park HS. The Orthodontic Treatment Using Micro-implant: The Clinical Application of MIA (Micro-implant Anchorage). Seoul: Narae, 2001.
45. Takano-Yamamoto T, Miyawaki S, Koyama I. Can implant orthodontics change the conventional orthodontic treatment? Dental Diamond 2002;27:26–47.
46. Tsunori M, Mashita M, Kasai K. Relationship between facial types and tooth and bone characteristics of the mandible obtained by CT scanning. Angle Orthod 1998;68:557–562.
47. Masumoto T, Hayashi I, Kawamura A, Tanaka K, Kasai K. Relationships among facial type, buccolingual molar inclination, and cortical bone thickness of the mandible. Eur J Orthod 2001;23:15–23.

Correction of Vertical Discrepancies

Facial esthetics is one of the main reasons patients seek orthodontic treatment. The allure of having a beautiful smile and an attractive profile drives many people toward treatments that often require significant individual sacrifice. Thus, the orthodontic experience should not merely be tooth alignment but a devoted effort to give the patient that entire new image he or she so desires. This chapter discusses some of the challenging vertical discrepancies commonly seen in orthodontic practice and the most effective methods of correcting them in an effort to enhance the quality of patients' lives.

Deep Bite Correction

The correction of a deep bite is essential to obtaining a harmonious facial profile and a balanced occlusion and must be addressed at the beginning of the leveling phase. Dental deep bite can be observed on both short- and long-faced patients, and its correction is necessary to properly reduce overjet (Fig 6-1a), especially in Class II, division 1 extraction cases. If deep bite correction is not considered during retraction, the palatal aspects of the maxillary incisors could interfere with the mandibular incisors, causing lingual tipping (Fig 6-1b), loss of anchorage, and temporomandibular joint (TMJ) dysfunction.

In Class II, division 2 deep bite cases, the maxillary incisors overlap the mandibular incisors excessively, "imprisoning" the mandible in the maxilla and forcing it to stay back. This tight relationship between anterior teeth limits mandibular function and prevents the jaw from growing normally. In growing individuals, it is imperative to correct maxillary incisor inclinations to make room (overjet) for the mandible to grow and develop normally.

6 | Correction of Vertical Discrepancies

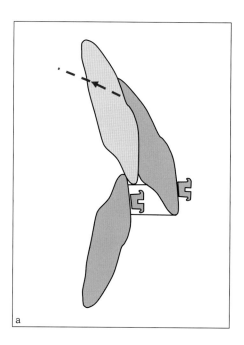 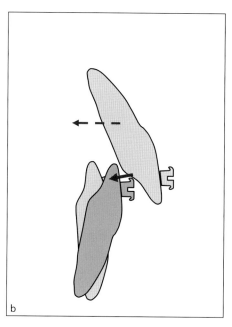

Fig 6-1 Class II, division 1 deep bite. *(a)* The deep bite should be corrected before or during maxillary incisor retraction to reduce the overjet properly. *(b)* During retraction, if the deep bite is not considered, the palatal surfaces of the maxillary incisors can interfere with the mandibular incisors, tipping them lingually.

Deep bite can be corrected in a number of ways, depending on the individual needs of the patient and the nature of the problem. Precise diagnosis and careful treatment planning are extremely important for a proper outcome. Some criteria such as vertical growth pattern, inclination of the occlusal planes, incisor-lip relationship, smile line, and vertical proportions of the face need to be considered before planning treatment.

In general, there are three methods of treating a deep bite:

- Intrusion of the incisors and extrusion of the posterior teeth
- Selective incisor intrusion
- Selective incisor extrusion[1-6]

A straight wire engaged in all brackets tends to intrude and protrude the incisors and extrude the molars. Figure 6-2 shows an occlusal view of a Class II, division 2 case with the straight wire causing incisor protrusion and molar expansion. In the sagittal plane, when the straight wire is engaged in brackets on incisors located below the maxillary occlusal plane, the overbite will be corrected mainly by incisor protrusion and intrusion (Fig 6-3). The maxillary occlusal plane inclination, however, may not be fully corrected because the canines and premolars will erupt as the incisors intrude and flare. If this had been a deep bite case combined with a gummy smile, the esthetic goals would not have been completely met because the smile line would not have been raised.

If incisor protrusion is indicated to correct Class II, division 2 malocclusion, then straight wire is the proper choice. However, some Class II, division 2 deep bite cases combined with severe crowding require extraction of maxillary premolars. In this instance, one should start treatment with canine retraction to make room for the crowded incisors, using a segmented arch to prevent flaring of the incisors and the problems associated with it, such as gingival recession and tendency to relapse. When the canines distalize, the incisors tend to move in the same direction by means of the transseptal fibers—which collectively form an interdental ligament that connects all the teeth of the arch—and the crowding often resolves spontaneously. Then brackets can be placed on the incisors and their intrusion can be accomplished with a continuous intrusion arch. After the deep bite is corrected, a straight wire can then be engaged to finalize leveling.

If one starts the treatment with a straight wire, the case will be converted from Class II, division 2 to Class II, division 1, owing to the incisor protrusion. The incisors must then be retracted to reduce the overjet. Protrusion followed by retraction may cause resorption of the in-

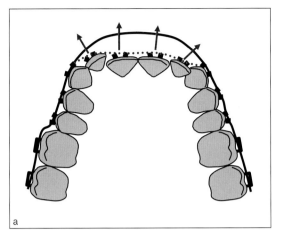
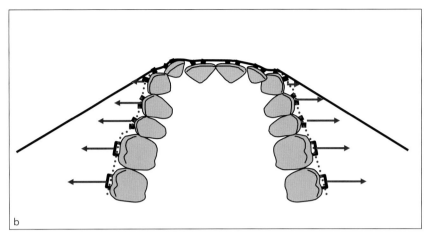

Fig 6-2 In a Class II, division 2 case, a straight arch engaged in the brackets would cause anterior protrusion *(a)* and posterior expansion *(b)*.

Fig 6-3 In the sagittal plane, when a straight wire is engaged in brackets on incisors erupted below the maxillary occlusal plane, overbite will be corrected by incisor protrusion. However, the maxillary occlusal plane inclination may not be fully corrected because the canines and premolars erupt as the incisors intrude and flare. At the completion of leveling *(red)*, the occlusal plane will be inclined downward anteriorly (α).

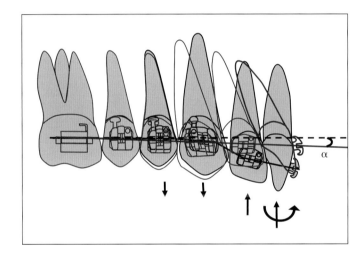

cisor roots from "round tripping" (jiggling). Therefore, straight wire should be avoided in Class II, division 2 cases when extraction is necessary.

During leveling with straight wire, tooth tipping might create premature contacts between antagonist teeth and cause the bite to open. In skeletal deep bite (low-angle) patients, bite opening with molar eruption is usually desired. In high-angle cases with deep bite, however, selective mechanics must be applied to control vertical growth of the face. Premature contacts in the posterior segment tend to stimulate clockwise rotation, resulting in an increase of lower facial height and worsening of the incisor-lip relationship and soft tissue profile. Occlusal forces normally cannot compensate for this bite opening because high-angle individuals have relatively weak chewing muscles. In growing patients, counterclockwise rotation can be achieved by intruding molars or controlling vertical movement of the posterior teeth and applying selective bite-opening mechanics, depending on the growth potential of the condyles.[7]

In normal- to low-angle patients, deep bite correction can be attained by any of the methods appropriate to the specific needs of the patient. In these cases, clockwise rotation of the mandible by molar extrusion is usually helpful in correcting the deep bite, increasing lower facial height, and improving the soft tissue profile. In skeletal deep bite (low-angle) patients, however, bite opening is a challenge. Even after having selective molar extrusion for a relatively long time, most patients undergo relapse because of their powerful chewing mus-

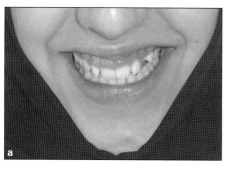
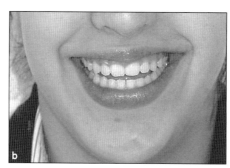
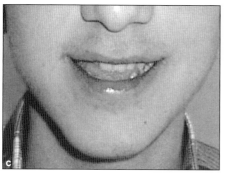
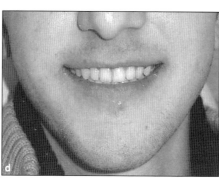

Fig 6-4 Changes in smile lines due to orthodontic treatment of a patient with a gummy smile *(a and b)* and a patient with a high smile line *(c and d)*.

cles and the nature of the malocclusion. These patients should be provided a fixed or removable anterior bite plane as a retainer to allow the molars to erupt.

Esthetics and the incisor-lip relationship

One of the main components of creating the esthetic smile coveted by patients is the correct alignment of the teeth, along with the harmony of this alignment with the lips (Fig 6-4). One thing that beautiful smiles have in common is that the teeth fill the labial commissures. The presence of dark corridors on each side of the maxillary arch during smiling gives the unpleasant effect of a narrow arch, while the appearance of the second premolar or molars is brighter and more esthetic.

The relationship between the maxillary anterior teeth and the lower lip is also an important factor affecting smile esthetics. Showing maxillary incisors during speaking or natural smiling is usually associated with a young and dynamic appearance. Showing the mandibular incisors—owing to the soft tissues' tendency to droop as a result of gravity—and hiding the maxillary incisors is, in contrast, regarded as older looking. It may therefore be necessary to intrude the mandibular incisors instead of the maxillary incisors for a more esthetic smile line in the long term.

Patients with a convex soft tissue profile associated with a vertical facial-growth pattern and retruded chin may also have a protrusive smile, even if maxillary protrusion does not exist. In these patients, lower facial height is usually increased by clockwise mandibular rotation; the lower lip strains and moves backward during smiling, giving the impression of maxillary protrusion (Fig 6-5).

The incisal edges of the maxillary incisors should follow the lower lip curve for an esthetic smile (Fig 6-6). In many deep bite patients, it may be necessary to open the bite by mandibular incisor intrusion only and level the maxillary anteriors so that the laterals are 0.5 mm gingival to the centrals.

Some patients with a deep bite also have a gummy smile. Treatment plans and final evaluation of these patients should not be based on radiographs and photos alone, which show only static relationships between the lips and incisors, but also on lip function.[8,9] Exposure of the maxillary incisors in relaxed lip position is the key to assessment of the natural smile.[10] One practical method that can be used to assess the smile line is to ask the patient to say the word *mamma* and then to keep the lips naturally apart. Peck et al[10] state that the average maxillary incisor exposure in rest position is 4.7 mm and 5.3 mm in 15-year-old boys and girls, respectively, com-

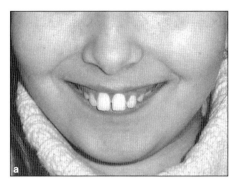
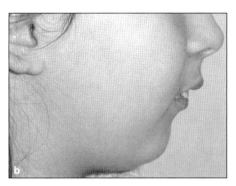

Fig 6-5 Protrusive smile (a) caused by a retruded chin (b).

Fig 6-6 For an esthetic smile, the incisal edges of the maxillary incisors should follow the lower lip curve.

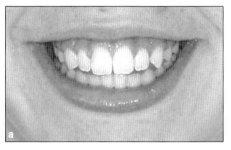
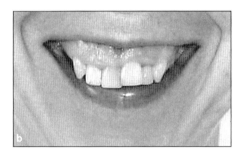
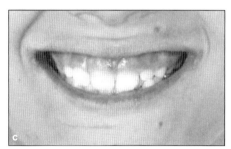

Fig 6-7 Gummy smiles due to (a) muscles raising the lip during smiling, (b) vertical overgrowth of the anterior maxilla, and (c) overeruption of maxillary incisors below the occlusal plane.

pared to an average of 9.8 mm and 10.5 mm in maximum smile, respectively. Dong et al[11] emphasize the effect of age on the smile line and show that the amount of incisor exposure, both at rest and during smile, drops dramatically with age. The average amount of tooth exposure at rest in individuals who are less than 30 years of age is close to 2 mm; in individuals who are 60 years or older, the average drops below 0. During a natural smile, it is normal to see as much as 2 to 2.5 mm of gingiva; anything beyond this is considered a gummy smile.

The smile can be affected by many factors, such as the muscles pulling the lip during smiling[10] (Fig 6-7a). However, the most common causes of a gummy smile are

- Vertical overgrowth of the anterior maxilla (Fig 6-7b)
- Overeruption of maxillary incisors below the occlusal plane (Fig 6-7c)

The first two causes are outside the limits of orthodontic treatment and must be evaluated within orthognathic and esthetic surgery concepts. Cases related to the third cause, however, can be treated orthodontically by selective incisor intrusion.

Certain patients present with gummy smile due to gingival hyperplasia, which can be confused with an orthodontic gummy smile. Therefore, differential diagnosis is imperative for proper treatment planning.

Selective incisor intrusion

The incisor-lip relationship is the best criterion for correcting a deep bite by selective incisor intrusion. When the maxillary incisors are positioned 2 to 3 mm below the upper lip and are below the maxillary occlusal plane, selective intrusion can be applied.[1,3,12] As previously noted, determination of the incisor-lip relationship should be done when the patient is at rest or speaking. Many adults show their mandibular incisors when speaking. These types of patients may require selective mandibular incisor intrusion. Incisor intrusion is also indicated in adults with severe periodontal bone loss. Some patients requiring selective incisor intrusion may

6 | Correction of Vertical Discrepancies

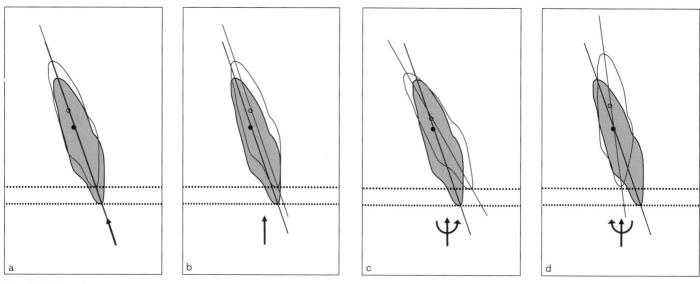

Fig 6-8 *(a to d)* Four approaches to intruding an incisor and the expected results *(red)*.

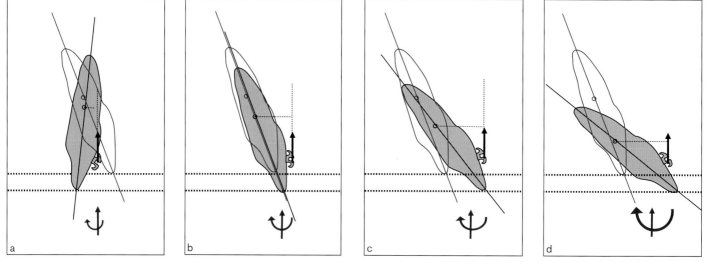

Fig 6-9 The type of intrusion of an incisor depends on its axial inclination. *(a to d)* As the axial inclination increases, so does the clockwise moment required during intrusion.

be high-angle cases, in which clockwise rotation of the mandible should be avoided. Mandibular incisor intrusion is easier in low-angle patients who present with a thick symphysis.[13] Incisor intrusion is also recommended for adults with periodontal bone loss.[14]

Clinically, intrusion is a difficult movement to achieve, and it requires three-dimensional controls. Incisor intrusion can be performed in several ways; Fig 6-8 shows four of these methods. For an incisor with normal inclination, bodily intrusion is not practical because the in-

trusive force should be a combination of vertical and horizontal force vectors, both of which must be kept under control throughout the movement (Figs 6-8a and 6-8b). Clinically, intrusion combined with some protrusion (Fig 6-8c) or retrusion (Fig 6-8d) would likely occur. In a typical Class II, division 2 malocclusion with upright and extruded maxillary incisors, intrusion combined with some protrusion would be beneficial to obtain a better inclination. Intrusion and retrusion is not a practical movement and not usually necessary because the apex of the root moves forward to a less favorable position.

Intrusion mechanics basically depend on the initial inclination of the incisor. Given the same amount of intrusive force applied to the bracket, the moment arm increases as the inclination of the tooth increases. Likewise, the more the inclination of the incisor increases, the more the clockwise moment to compensate for it increases (Fig 6-9).

Clinically, continuous intrusion arches using the segmented arch technique of 2 × 4 archwires (2 × 4s) work relatively well in selective incisor intrusion.

Continuous intrusion arch

A continuous intrusion arch is bent from a 0.018 × 0.025–inch stainless steel (SS) wire with a 2.5-mm helix or a plain 0.017 × 0.025–inch titanium-molybdenum alloy (TMA) wire (Fig 6-10). Continuous intrusion arch mechanics are similar to those of 2 × 4s. A sweep (V) bend is placed in the wire at the level of the first premolars (or first primary molars). If incisor protrusion is indicated, the arch is ligated on the central incisor brackets or at the midline.[15] For incisor protrusion, the arch should slide easily through the molar tubes. With these mechanics, the deep bite will be corrected with both incisor protrusion and intrusion.[16–18] This is especially useful in correcting Class II, division 2 deep bite cases with mandibular retrusion in growing patients. As the maxillary incisors protrude, the mandible will have more room to grow forward to reduce the overjet and reach a Class I relationship.

If bodily intrusion of the incisors is necessary, the line of action of the intrusive force should pass through the center of resistance of the four incisors via ligation of the intrusion arch to the main archwire at the distal wings of the lateral incisor brackets. Clinically, pure bodily intrusion is difficult owing to the complexity of the movement. A slight change in the relationship of the line of action of the force with the center of resistance can change the type of movement. If the force passes anterior to the center of resistance, the incisors protrude, which can be prevented with a light chain elastic.

Correction of deep bite with segmented arch mechanics requires initial leveling and alignment of both anterior and posterior segments separately, which takes time. The leveling can be accomplished with a 0.014-

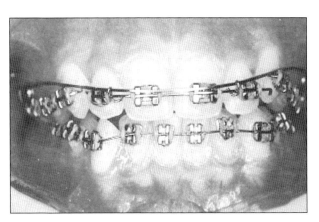

Fig 6-10 Frontal view of a continuous intrusion arch ligated to the anterior teeth between the central and lateral incisors. Note the shape of the arch.

inch nickel titanium (NiTi) wire in combination with a continuous intrusion arch, making the mechanics more efficient and shortening chair time. A 0.014-inch NiTi wire is a very flexible wire that has almost no negative effect on anchorage teeth, yet it can efficiently correct incisor crowding. The 0.017 × 0.025–inch TMA or 0.018 × 0.025–inch SS wire can control the occlusal plane if it is used in combination with a 0.014-inch NiTi wire (Figs 6-11 and 6-12). The continuous intrusion arch can be ligated to the lateral incisor brackets to hold those teeth in place while the NiTi wire intrudes the central incisors.

If the intrusive force is light, the reactive extrusive force on the molars is also light and therefore kept in check by the forces of occlusion. Table 6-1 shows the optimum intrusive forces for maxillary and mandibular anterior teeth.[2] In high-angle cases, molar extrusion may need to be actively prevented to avoid clockwise rotation of the mandible. In these cases, high-pull headgear with long arms combined with a transpalatal arch works relatively well (see Figs 5-7 and 6-35).

Microimplant mechanics

Microimplants (temporary anchorage devices [TADs]) provide good anchorage support for selective incisor intrusion in both the maxilla and mandible. The microimplant is usually placed at the midline between the central incisor roots. Intrusion force can be applied directly on the archwire if some protrusion is also indi-

6 | Correction of Vertical Discrepancies

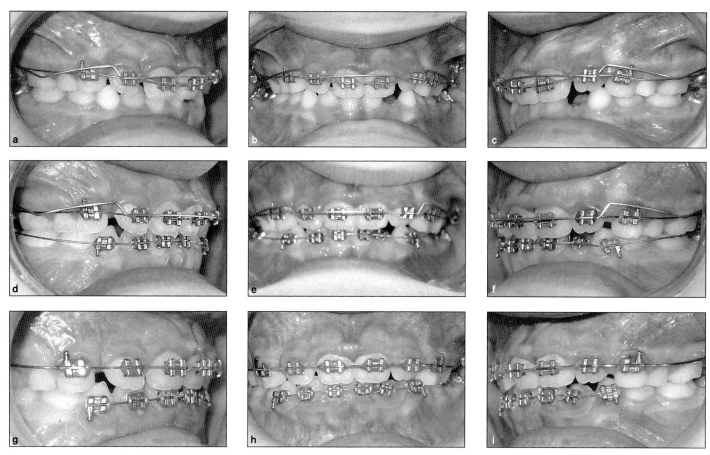

Fig 6-11 Correction of deep bite can be accomplished with a 0.014-inch NiTi wire combined with a continuous intrusion arch. *(a to c)* Beginning of treatment. The intrusion arch is tied on the lateral incisors, while NiTi wire is inserted into the central incisor brackets. *(d to f)* After 4 weeks, the central incisors have been intruded to the level of the lateral incisors. *(g to i)* After 12 weeks of treatment, a 0.016 × 0.022–inch SS wire is engaged in the brackets.

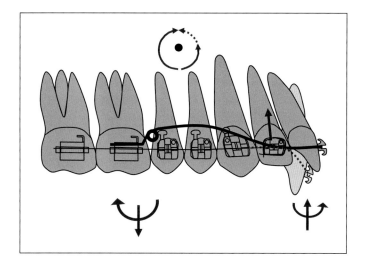

Fig 6-12 A continuous intrusion arch can be used in combination with a straight leveling wire. This arch can be ligated to the lateral incisor brackets so that the line of action of the force passes close to the center of resistance of the incisors. The central incisors intrude and protrude with the straight wire as the lateral incisors keep their positions with the effect of the intrusion arch.

cated (Fig 6-13). If protrusion is not desired, the archwire can be cinched back so the teeth can be intruded more effectively. In severe deep bite cases, two TADs can be inserted between the central and lateral incisors to obtain more effective intrusion and to correct transverse asymmetry, if present (Fig 6-14).

Deep Bite Correction

Table 6-1	Recommended intrusive forces and headgear for deep bite correction*	
Teeth to be intruded	**Force (g) per side**	**Headgear**
Maxillary central incisors	15–20	Occipital-Ant to CR
Maxillary central and lateral incisors	30–40	Occipital-Ant to CR
Maxillary central and lateral incisors and canines	60	Occipital-Ant to CR
Mandibular central incisors	12.5	Cervical-Ant to CR
Mandibular central and lateral incisors	25	Cervical-Ant to CR
Mandibular central and lateral incisors and canines	50	Occipital-Ant to CR
Mandibular canines	25	None

*Reprinted from Burstone[2] with permission.
Ant = anterior; CR = center of resistance.

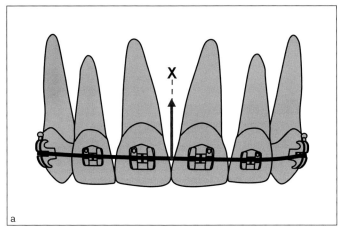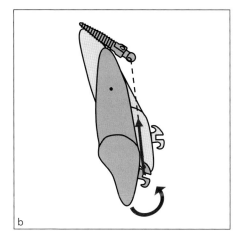

Fig 6-13 A microimplant, or TAD, can be used to correct anterior deep bite as well as the cant of occlusal planes. (*a*) The TAD is inserted between the incisors, and the force is applied to the archwire to obtain selective incisor intrusion. (*b*) In this technique, some incisor protrusion is also expected.

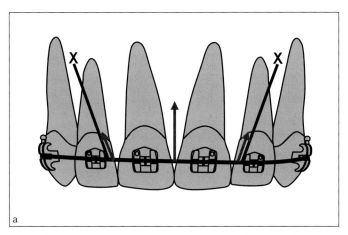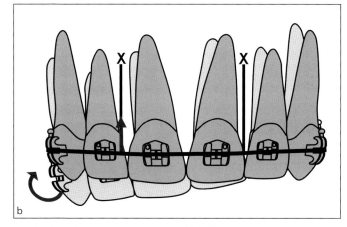

Fig 6-14 In patients with severe deep bite, two TADs can be inserted between the lateral incisors and canines (*a*). TADs can also be used to correct the transverse cant of the maxillary occlusal plane (*b*).

6 | Correction of Vertical Discrepancies

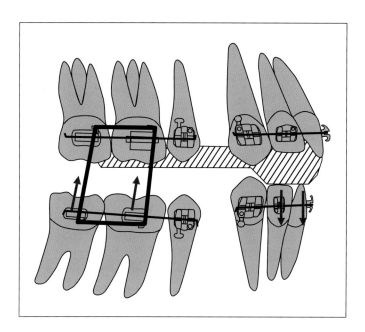

Fig 6-15 Deep bite correction by selective molar extrusion. Mandibular molars extrude with the effect of vertical elastic, while the anterior bite plane raises the bite and holds the maxillary molars in place. Because of reciprocal anchorage, the mandibular molars move up as the mandibular incisors tend to move downward and forward.

Selective molar extrusion

In deep bite cases wherein lower facial height is reduced and the maxillary incisor–lip relationship is good, correction of the deep bite should be accomplished with selective molar extrusion. In these cases, bite opening occurs mainly through clockwise mandibular rotation, with the chin moving downward and backward to improve the soft tissue profile.

Using an anterior bite plane or bite raiser in combination with posterior vertical elastics is an efficient way to extrude molars (Fig 6-15). The anterior bite plane is a removable appliance used on the maxillary dental arch in combination with a fixed appliance. Two C-clasps at the molar area hold the appliance in place. The anterior bite plane keeps the bite raised, while posterior elastics are used to extrude the molars. If mandibular molar extrusion is indicated (Class II cases), these teeth can be incorporated into a segmented arch, or a light, straight leveling wire can be engaged in the brackets. Because the acrylic plate covers the occlusal surfaces of the maxillary molars, they are not affected by the elastic force.

Because the bite plane is removable, it needs to be removed for meals. However, even in this short period, the chewing forces, which are relatively high in low-angle patients, could reintrude the extruded molars and reduce the efficiency of the mechanics. Also, whereas the reaction of the extrusive elastics tends to intrude the incisors, molar extrusion should be obtained as soon as possible to avoid excessive incisor intrusion or protrusion. Fixed anterior bite raisers should therefore be used to produce a more effective bite opening. Figure 6-16 shows a patient being treated with anterior fixed bite raisers in combination with posterior zigzag elastics to extrude the molars. Bite raisers and elastics were placed right after bracketing in combination with 0.014-inch NiTi leveling wires. With this method, the bite opened dramatically within 2 months, with no significant change in the maxillary smile line.

In treating skeletal Class II patients, cervical headgear can be used to correct both the molar relationship and the deep bite. The long and upward-angulated arms of this type of headgear are effective in opening the bite by molar distalization and extrusion[19,20] (see Fig 5-5a).

Molar extrusion cases have a strong tendency to relapse after treatment because of the relatively high occlusal forces of low-angle patients and stretching of the supra-alveolar gingival tissue.[21] Retention of these cases should be performed with either a removable retainer and anterior bite plane or maxillary and mandibular fixed lingual retainers combined with anterior bite raisers to allow the molars to move occlusally.

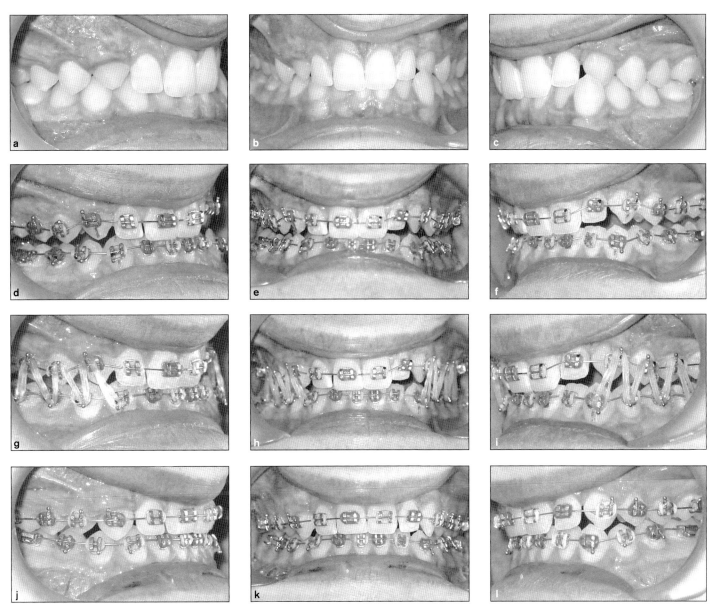

Fig 6-16 Deep bite correction with selective molar extrusion mechanics. *(a to c)* Before treatment. *(d to i)* Bite raisers were bonded to the palatal aspects of the maxillary incisors, and up-and-down elastics, which were worn only at night, were placed to extrude the molars. *(j to l)* A dramatic change in the bite was achieved in 2 months.

Correction of the curve of Spee

The curve of Spee is an occlusal characteristic of certain deep bite patients. It is a common understanding among clinicians that this curve should be flattened to obtain a more functional occlusion, yet it is debatable whether the curve should be corrected in all cases.[22]

Treatment plans are usually made with the belief that space is needed in proportion to its depth to correct the curve of Spee, but this is a fallacy.[22] In deep bite cases, it is important to make a differential diagnosis regarding whether the curve of Spee is stepped or angulated before engaging wires to flatten the curve.

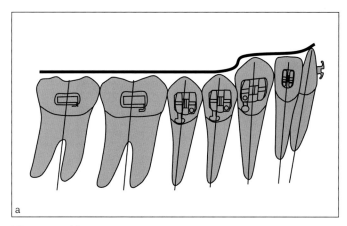 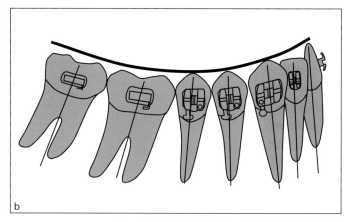

Fig 6-17 *(a)* In the stepped curve of Spee, the axial inclinations of the teeth are usually parallel. In these cases, steps exist at the first premolar–canine and canine–lateral incisor contacts, so space is not needed to flatten the curve. Intrusion of the incisors and canines, or extrusion of the molars only, might eliminate the problem. *(b)* In an angulated curve of Spee, the mandibular teeth are aligned on a curved arch form, so space is needed to flatten the arch.

Stepped curve of Spee

In a stepped curve of Spee, the axial inclinations of the teeth are usually satisfactorily parallel to one another, but there may be steps between the posterior and anterior teeth that resemble a curve (Fig 6-17a). This can be observed in some mandibular retrusion cases in which the mandibular incisors are overerupted from lack of contact with antagonists. Because this is typically a deep bite case with a mandibular curve of Spee, one tends to correct it using straight or reverse-curved wires. In this case, a differential diagnosis is necessary to make a proper treatment plan. The correction of the stepped curve of Spee is achieved with selective incisor intrusion or molar extrusion, depending on the vertical growth pattern of the patient. Because the teeth are almost parallel to each other, there is no need to expand the arch either anteroposteriorly or transversely to gain space. Engagement of a straight wire in the brackets will result in incisor intrusion and protrusion along with premolar extrusion. A straight wire will level and align the dental arch perfectly, but the occlusal plane inclination and incisor-lip relationship may not be completely corrected.

Angulated curve of Spee

In the angulated curve of Spee, space is needed to level the teeth in as much as they are aligned on a curve (Fig 6-17b). If a flexible straight or reverse-curved wire is engaged in the brackets, the incisors will intrude and flare, the molars will tip backward, and the premolars will erupt. In many cases, this helps correct the deep bite by premolar extrusion and incisor protrusion. If incisor protrusion or molar tipback is not desired, a differential diagnosis should be made to avoid any adverse effects.

Use of straight or reverse-curved archwires

Straight or reverse-curved wires are commonly used in the correction of deep bite with a curve of Spee. Use of these wires results in incisor protrusion, molar tipback, and premolar extrusion (Fig 6-18a). Occlusally, the reverse-curved archwire should be slightly widened at the premolar area to compensate for any palatal tipping during extrusion and to avoid crossbite (Fig 6-18b). This technique works quickly and efficiently because extrusion and tipping are the easiest movements. Molar tipback and incisor protrusion provide room for the premolars by increasing the anteroposterior dimensions of the arch. Figure 6-19 shows an anterior deep bite case in which a reverse-curved wire is indicated. In cases requiring selective incisor intrusion or selective molar extrusion, however, straight-wire mechanics are not suitable. In such cases, differential mechanics should be used to obtain proper occlusal plane inclination and better incisor-lip relationship and to avoid clockwise rotation of the mandible. Furthermore, in many patients, incisor protrusion should be avoided because of periodontal or stability concerns.

Deep Bite Correction

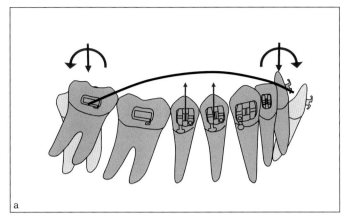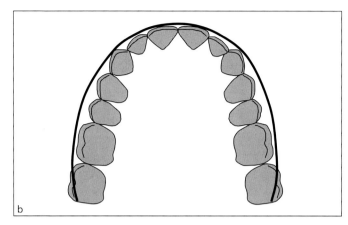

Fig 6-18 *(a)* Application of a reverse-curved wire to the mandibular arch results in incisor protrusion and intrusion, molar tipback and intrusion, and premolar extrusion. *(b)* Occlusally, the reverse-curved archwire should be slightly widened at the premolar area to compensate for palatal tipping during extrusion and to avoid crossbite.

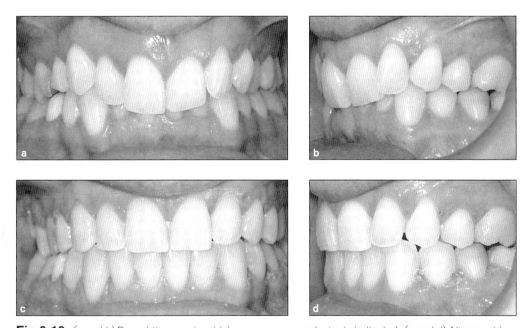

Fig 6-19 *(a and b)* Deep bite case in which a reverse-curved wire is indicated. *(c and d)* Alignment has been done with maxillary incisor protrusion and intrusion as well as premolar extrusion.

Differential mechanics

As explained previously, once a straight wire is engaged in the brackets, all movement takes place simultaneously. If incisor protrusion needs to be avoided and space needs to be gained by distal tipping (ie, to upright) of the molars, then differential mechanics must be applied, as shown in Fig 6-20. A segmented leveling archwire can be engaged from canine to canine to avoid protrusion. A cantilever from a 0.017 × 0.025–inch TMA or 0.018 × 0.025–inch SS wire with a 2.5-mm helix is engaged in the molar tube. The mesial end should be placed on the anterior wire between the lateral incisor and canine to avoid anterior protrusion. As the molar uprights and tips distally, room is created for extrusion of the premolars. At this point, a flexible straight wire can be placed in the brackets for leveling.

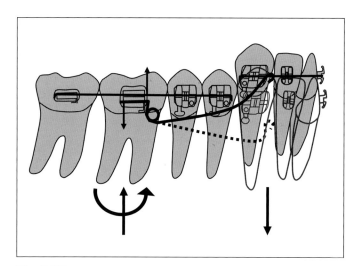

Fig 6-20 In an angulated curve of Spee case, if incisor protrusion needs to be avoided, differential mechanics should be applied to provide room in the dental arch. A cantilever between the molars and anterior teeth can be used to upright the molars and gain space without protruding the incisors.

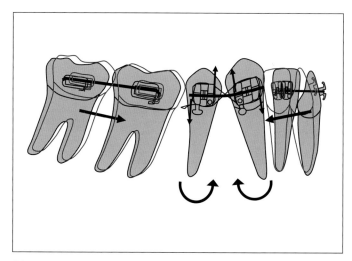

Fig 6-21 In an angulated curve of Spee case, if extraction is indicated and if incisor protrusion should be avoided, a segmented archwire can be placed initially in the canine and premolar brackets to upright roots and close the space. As the space closes, the incisors and molars will drift naturally toward the extraction space. A straight wire can then be engaged in the brackets for general leveling and alignment.

If extraction is indicated, a straight wire placed in the brackets will cause the premolar and canine crowns to move away from each other, enlarging the extraction space. In addition, the incisors will flare and the molars will tip back. Retraction of the already protruded incisors may result in so-called round tripping and root resorption. In this case, differential mechanics can be applied to save time and avoid adverse effects. First, a segmented, flexible wire is engaged in the premolar and canine brackets with a figure-eight ligature to prevent the crowns from separating and to obtain root movement (Fig 6-21). Once the roots are corrected, the extraction space can be closed with a chain elastic. Room must be made for the incisors and molars, which tend to drift toward the extraction site. At this point, a straight wire is placed, and leveling can be done without incisor protrusion.

Class II, division 2 cases

Correction of Class II, division 2 cases can be accomplished in one of two ways, depending on whether the patient is fully grown.

In growing patients, mandibular growth is usually prevented by a deep overbite because of the inclination of the maxillary incisors. It is therefore necessary to correct the overbite by maxillary incisor intrusion and protrusion and to increase the overjet to allow the mandible to grow normally. Although mandibular growth is basically dependent on condylar activity, it can be stimulated by functional appliances such as activators and fixed bite jumpers.

Burstone[23] has suggested that correction of the Class II relationship is associated with maxillary molar distalization, which occurs as a reaction to maxillary incisor protraction and upper lip forces. He also suggested that in the long term, the mandible would grow normally, which is expected in all Class II cases. It is a clinical fact that removal of the obstacle in front of the mandible is helpful in correcting anteroposterior as well as vertical relationships of the jaws and the lips.

In fully grown individuals, extraction or stripping of the arch to gain space is a common treatment strategy unless a surgical option is preferred as a solution for the deep bite and mandibular retrusion. Differential mechanics should be used to avoid possible adverse effects from the use of straight wire. In extraction cases, canines should be distalized individually, first with a segmented archwire to manage the anterior crowding. After canine distalization, the incisors can be aligned and intruded

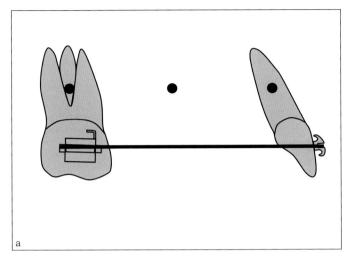 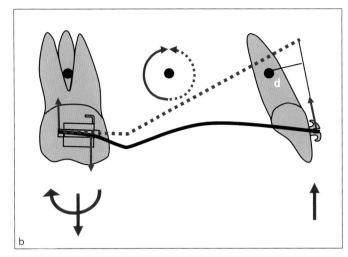

Fig 6-22 *(a)* If a round, straight wire passes through the molar tubes and incisor brackets located on the same plane in a 2 × 4 system, no movement will be observed since there is no force on the system. *(b)* When a V-bend is placed in the wire, a force system is created as soon as the wire is placed in the brackets. The clockwise moment on the molar is balanced with intrusive and extrusive forces, turning the whole system in a counterclockwise direction.

with a continuous intrusion arch, then a straight wire can be engaged in the brackets to finalize the alignment and correct incisor inclination.

If a straight, flexible wire is used initially, round tripping of the incisors would be expected owing to the protrusion and retraction movements. In addition, the maxillary incisor–lip relationship and smile line would not have been corrected properly because of the improper inclination of the maxillary occlusal plane (see Fig 6-3).

Transitional dentition cases

Deep bite in the early or late transitional dentition must be corrected because of the reasons already mentioned. Several options exist using either removable or fixed appliances. A maxillary removable appliance with anterior bite plane, combined with maxillary push springs, can correct the bite with both incisor protrusion and molar extrusion. Fixed options such as 2 × 4 mechanics are also available to extrude and upright the molars, as well as to intrude the incisors with more precise alignment than removable appliances would provide.

2 × 4 arch mechanics

The 2 × 4 arch includes two molars and four incisors. The Ricketts utility arch, Burstone continuous intrusion arch, and straight wires with sweep or V-bends are examples of 2 × 4s. Even though there are some differences in application or shape of the wires, the mechanics are similar.

In 2 × 4 mechanics, because of the long distance between anterior and posterior segments, the force system obtained is optimum and long-ranged. If the wire is flexible enough, incisor crowding can be corrected along with the deep bite.

Figure 6-22a shows an example of a 2 × 4 from 0.016-inch SS. When this wire passes through the molar tubes and incisor brackets located in the same plane, no tooth movement will be observed because there is no deflection (force) of the wire. In this force system, consisting of two molars and four incisors, anchorage values of the anterior and posterior segments are assumed to be equal, so the center of resistance of the system is in the middle. When a V-bend is made on this wire close to the molar, a force system is created (Fig 6-22b). This force system works according to the principles of V-bend mechanics (ie, the position of the V-bend between attachments). A clockwise moment occurs on the molar, and this moment is balanced with intrusive and extrusive balancing vertical forces on the incisors and molars, respectively. Because the wire is round and does not have third-order interaction within the incisor bracket slot, it can be assumed to act as a cantilever. In other words, it applies force at a single point. Note that the intrusive

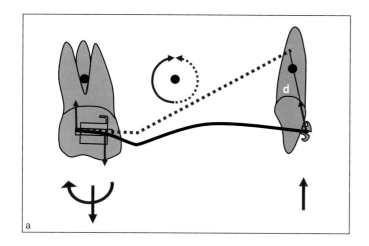
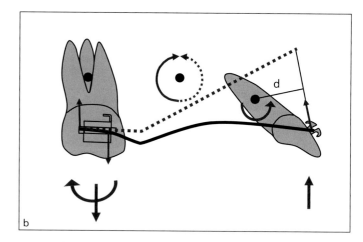
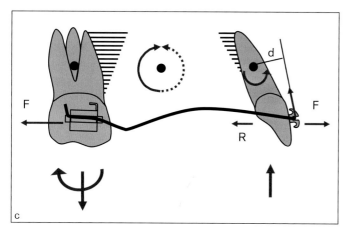

Fig 6-23 The anterior balancing force tends to intrude the incisor. *(a)* If this force passes through the center of resistance, it intrudes the tooth with translation. *(b)* In an excessively protrusive incisor, the line of action of the force passes far labial to the center of resistance and creates a high moment. *(c)* If the wire is cinched back, a tug-of-war is initiated between the anterior and posterior segments because they tend to move in opposite directions. Practically, the molar with the higher moment and anchorage either prevents the incisor from moving forward or pulls it backward.

force in the anterior passes labial to the center of resistance of the incisors and causes a counterclockwise moment. This is a moment of force and not a couple. If this force passes through the center of resistance, it causes bodily intrusion (Fig 6-23a). The amount of this moment increases as the inclination of the incisor increases (Fig 6-23b). As explained in the section "Mechanics of V-Bend Arches" in chapter 3, the location of the V-bend has considerable effect on the force system and final positions of the teeth. As the V-bend approaches the molar, the clockwise moment increases, as do the balancing intrusive and extrusive forces.

If the same arch is cinched back, a tug-of-war is initiated between the anterior and posterior segments owing to their opposite moments.[24] Clinically, molars with greater moments and anchorage either prevent the incisors from moving forward or pull them backward (Fig 6-23c). During this process, the molar crowns tip distally while the roots move mesially.[25]

If the round wire is replaced with rectangular wire, the force system changes because the rectangular wire interacts with the incisor bracket. The force system is associated with the balance between the degrees of tipback angulation (posterior moment) on the molar and the torque angulation (anterior moment) on the incisor.

If the clockwise moment on the molar and counterclockwise moment on the incisor are assumed to be equal, the system reaches static equilibrium with protrusion of the incisors and distal tipping of molars, with no vertical balancing forces (Fig 6-24a). If the wire is cinched back, the crown movements stop, but because the moments are still acting on the teeth, the molar roots move mesially and the incisor roots move palatally.

If the tipback angle (posterior moment) is increased, the system reaches static equilibrium with vertical balancing forces, which are intrusive at the incisors and extrusive at the molars (Fig 6-24b). The amount of intrusive and extrusive forces depends on the amount of

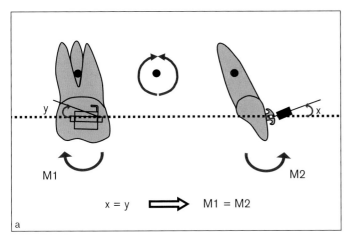
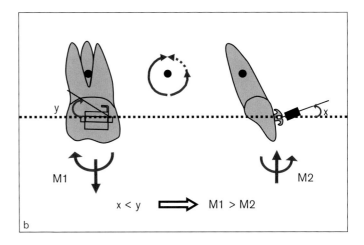
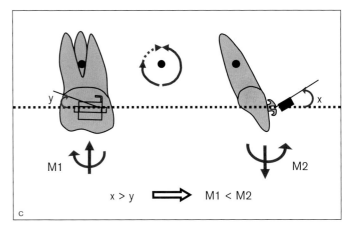

Fig 6-24 The force system changes if a rectangular wire is used instead of a round wire. *(a)* If the tipback angle (y) is equal to the torque angle (x), equal and opposite moments are obtained on both ends. *(b)* If the tipback angle is higher than the torque angle, the system reaches static equilibrium with intrusive force on the incisor and extrusive force on the molar. *(c)* If the torque angle is higher than the tipback angle, the system reaches static equilibrium with extrusive force on the incisor and intrusive force on the molar, which results in incisor intrusion and deepening of the bite.

tipback (clockwise moment) on the molar. Clinically, optimum intrusive force can be obtained only if the moment on the molar is optimum.

If the torque angle (anterior moment) is increased, the system reaches static equilibrium with vertical balancing forces, which are intrusive at the molar and extrusive at the incisor (Fig 6-24c). Clinically, this maneuver results in overeruption of the incisors and severe deep bite because molar intrusion does not occur easily.

Utility arch

The Ricketts utility arch is an example of a 2 × 4 that can be used for various purposes, both at the beginning and at later stages of treatment. The advantage of the utility arch is that it provides effective control over the incisors' and molars' axial inclinations, sizes of the dental arch, and vertical relationship between maxillary and mandibular jaws during the transitional dentition.[13,25]

The utility arch consists of five main sections (Fig 6-25): the anterior section (A), anterior step (B), buccal bridge (C), posterior step (D), and posterior section (E). A classic utility arch is formed from 0.016 × 0.016–inch Elgiloy wire (Rocky Mountain Orthodontics) (in 0.018 slots). The anterior section is formed first, and then the anterior and posterior steps are bent. Bending the anterior step at more than 90 degrees will later help prevent the arch from irritating the gums when the incisors intrude. The height of the steps must be around 3 to 5 mm, depending on the case.

The buccal bridge should pass 1 mm away from the attached gingiva. When the utility arch is engaged in the brackets and tubes, the distal step should contact the mesial of the molar tube. The utility arch causes tipback and extrusion of the molars and protrusion and intrusion of the incisors (Fig 6-26a). Incisor protrusion can be prevented by either cinching back or placing lingual root torque in the mandibular incisors. Labial root

6 | Correction of Vertical Discrepancies

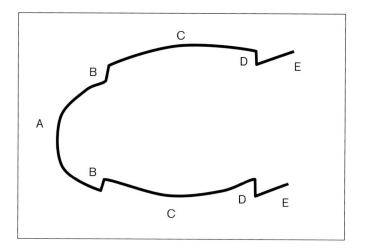

Fig 6-25 The sections of the (maxillary) utility arch: anterior section (A), anterior step (B), buccal bridge (C), posterior step (D), and posterior section (E).

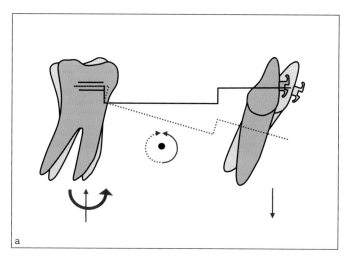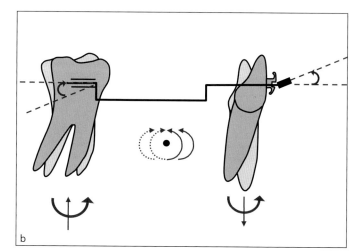

Fig 6-26 *(a)* With the effect of tipback bend, the molars upright and extrude while the incisors protrude and intrude. *(b)* Labial root torque of 5 to 10 degrees prevents protrusion of the incisors, increasing the intrusive effect.

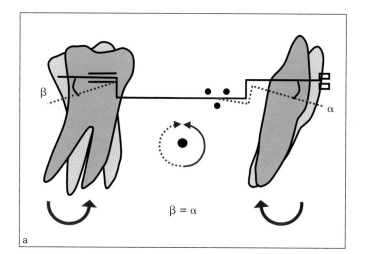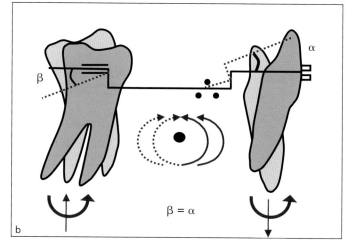

Fig 6-27 Intraoral activation of the utility arch. *(a)* Lingual root and labial crown movement on the incisor and distal tipping on the molar. If vertical balancing forces need to be avoided, it is important that the molar tipback angle (β) and anterior V-bend angle (α) are equal. *(b)* Labial root torque on the incisors creates same-directional moments, resulting in vertical balancing forces.

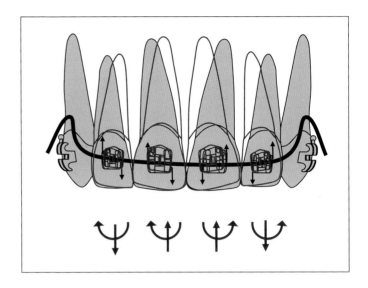

Fig 6-28 If a 2 × 4 archwire is engaged directly in the incisor brackets, the roots tend to move toward the midline. To avoid this adverse effect, the tipback bend should be kept as low as possible. (Reprinted from Burstone[1] with permission.)

torque of 5 to 10 degrees increases the intrusive effect on the incisors and the extrusive effect on the molars. Another reason for this torque is to keep the mandibular incisors in trabecular bone during intrusion and to prevent their roots from contacting the lingual cortical bone[13,25] (Fig 6-26b).

The utility arch can also be used to reinforce molar anchorage by moving the roots into buccal cortical bone and giving them buccal root torque. As the molars extrude with the tipback, they tend to tip lingually, but enlarging the arch buccally can compensate for this tendency. Toe-in bends of 30 to 45 degrees prevent the mesiolingual rotation of molars stemming from mesially directed Class II elastics and help achieve a Class I relationship in the posterior segment.[13,25]

If further activation is required, it can be accomplished intraorally with a three-jaw or Tweed pliers. If vertical balancing forces or transverse asymmetry are not desired, it is important to put an equal amount of activation on both sides of the arch (Fig 6-27).

Utility arch versus segmented intrusion arch mechanics

In the sagittal plane

The utility arch and the segmented intrusion arch present similar functional characteristics, even though they have different shapes. The main difference between these two bite-opening methods is the manner in which the force is transferred to the incisors. The utility arch (or any 2 × 4 continuous arch with a tipback or curve of Spee) is engaged in the incisor brackets, which usually results in protrusion and intrusion of the incisors and tipback and extrusion of the molars. Preventing incisor protrusion is possible only by cinching back the archwire or by putting labial root torque on the incisors. The segmented intrusion arch, however, is not directly engaged in the brackets (see Figs 6-11 and 6-12). Therefore, it does not apply a couple on the incisors, but it does on the molars. To obtain incisor protrusion, the arch should be ligated to the base arch at the labial aspects of the incisors. To avoid protrusion and obtain bodily intrusion, it is tied at the distal wing of the lateral incisor. The segmented intrusion technique provides more movement control by adjusting the point of application of the force.

In the frontal plane

In the frontal plane, when a utility (or a continuous intrusion) arch is deflected and engaged in the incisor brackets, it tends to move the roots toward the midline (Fig 6-28). This side effect is more pronounced as the tipback angle increases and can be avoided only by keeping the tipback angle low (around 5 degrees). In segmented intrusion mechanics, this side effect does not occur because the wire is not fully engaged in the brackets.

Treatment of High-Angle Cases and Correction of Open Bite

Open bite can be either dental, when it is limited to one or a couple of teeth, or skeletal, when it is rooted in the maxillary or mandibular basal bones. Dental open bite can be seen in all facial growth types, including low angle, because the opening is usually caused by habits such as thumbsucking, tongue thrusting, infantile swallowing, nail biting, inserting objects between the teeth, or a more functional activity such as mouth breathing. Skeletal open bite, however, is a more complicated and multifactorial anomaly. Skeletal open bites are usually associated with a high-angle growth pattern and increased lower facial height. Habits such as tongue thrusting are usually secondary in these cases, and they often stem from the tongue's adaptation to the opening between maxillary and mandibular incisors. Treatment of open bite, then, depends on the nature of the malocclusion.

Correction of dental open bite

A habit-breaking appliance can be used to overcome the behaviors that most commonly cause this malocclusion, but, as previously stated, these habits are often functional in origin and thus more difficult to control. Children (usually in the transitional dentition) with respiratory airway resistance, for example, tend to hold their tongues over their mandibular anteriors to create an airway.[26] Before orthodontic treatment, consultation with an ear, nose, and throat specialist to clear the nasal airway is advisable in such a case. For open bites caused by thumbsucking or other deleterious habits, an interfering appliance such as a tongue crib, in combination with maxillary and mandibular 2×4 mechanics, can be used to close the bite and perhaps eliminate the behavior that opened it. Lip and tongue exercises that train the tongue muscle and strengthen the surrounding musculature, used in conjunction with the crib, are also important to ensure long-term stability.

Correction of skeletal open bite

Besides a vertical growth pattern, factors such as adenoid vegetation, polyps, tumors, septum deviation, and narrow nostrils that block the nasal airway play an important role in the etiology of skeletal open bite. A patient with a blocked nasal airway must keep his or her mouth open while the tongue stands on the mouth floor with slightly anterior position to keep the air passage open for breathing.[26] This tongue position prevents the anterior teeth from erupting, or at least intrudes them while the posterior teeth erupt freely. Extrusion of the molars results in clockwise rotation of the mandible and an increase in lower facial height. This increase becomes worse and results in a skeletal open bite if it is not compensated for by condylar adaptation.

Constriction of the maxillary arch is a common symptom in mouth breathers, which makes this picture more complicated. In these cases, the low and anterior tongue position permits the external musculature to constrict the unsupported maxillary dentoalveolar bone. This, in turn, may cause posterior crossbite, premature molar contact, and increased facial height by clockwise mandibular rotation.

Diagnosis and treatment planning of the skeletal open bite should be made carefully by considering all these functional factors as well as any habits that might cause deformation of the bone. Cephalometric evaluation is also important for proper diagnosis. In a study of 50 patients to determine the relationship between anterior open bite and the vertical growth pattern, a high correlation was observed between the mandibular plane angle and the Jarabak ratio (PFH/AFH).[27] In most of the patients with dental anterior open bite, the mandibular plane angle, Jarabak ratio, and palatomandibular plane angle were observed to be above the normal limits, indicating a skeletal open bite. Mandibular plane angle, Jarabak ratio, and palatomandibular plane angle values indicated that most patients diagnosed with a dental open bite actually had a skeletal open bite.

Some patients with skeletal open bite present with canting of the maxillary occlusal plane, which is usually associated with a high smile line. Typically, when one of these patients smiles, a dark area is seen between the lips. The upper lip usually hides the maxillary incisors, either partially or completely (Fig 6-29). In this case, the goal of treatment must be to restore the lip-incisor relationship by correcting the cant of the maxillary occlusal plane with selective maxillary incisor extrusion. Unlike intrusion, incisor extrusion is not difficult to achieve,

Fig 6-29 The smile line is affected by the cant of the occlusal plane. This patient's high smile line (a) needed to be restored with maxillary incisor extrusion (b).

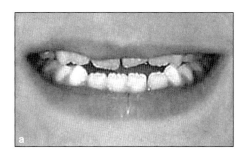
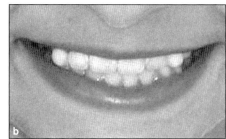

but it requires careful mechanics providing three-dimensional control. Incisor extrusion can be performed in several ways; Fig 6-30 shows four possible methods. Bodily extrusion of an incisor can be achieved with a combination of vertical and horizontal vectors that should be controlled throughout tooth movement (Figs 6-30a and 6-30b). Extrusion with controlled tipping (Fig 6-30c) can be obtained with a combination of vertical force and palatal root torque. However, palatal root torque is not clinically feasible for such high incisors. Extrusion with uncontrolled tipping is shown in Fig 6-30d. A straight wire passing through the incisor brackets and anterior up-and-down elastic force causes extrusion with either controlled or uncontrolled palatal tipping, depending on the amount of extrusion.

The type of extrusion depends on the inclination of the incisors. Vertical elastics applied to the palatal button provide a more favorable force vector, causing bodily extrusion or a combination of extrusion and protrusion, depending on the direction of pull (Fig 6-31a). For a vertical force applied on the bracket, the more the tooth is protruded, the more it tips in an uncontrolled manner because of the moment arm between the center of resistance and the line of action of the force (Figs 6-31b to 6-31d). Clinically, a flexible straight wire helps level the maxillary teeth by incisor extrusion, but this is usually combined with slight palatal tipping that can cause labial root movement. Additionally, the canines and premolars may intrude as a reaction to extrusion, which makes leveling more complicated. Up-and-down elastics in the premolar-canine area after full mandibular leveling help stabilize the posterior segment and provide good anchorage for segmented arch mechanics, although vertical elastics on flexible wires may cause the mandibular teeth to extrude and change the mandibular occlusal plane.

The segmented arch approach provides more predictable and favorable mechanics than straight wires in the correction of anterior open bite with incisor extrusion. The maxillary anterior and posterior segments can be leveled separately using up to 0.017×0.025–inch SS wires, and a 0.016×0.022–inch TMA wire can be engaged as an auxiliary continuous extrusion arch. The possible counterclockwise rotation of the maxillary posterior segment can be prevented by using up-and-down elastics on the canines and premolars. Leveling the mandibular arch up to a rectangular wire might minimize or even eliminate the reaction. The TMA extrusion arch should be tied at the distal wings of the lateral incisors to obtain more bodily extrusion (Fig 6-32a). A microimplant inserted between the maxillary first molar and second premolar provides indirect anchorage to extrude the incisors by means of a V-shaped bar (Fig 6-32b). This bar is soldered to crimpable tubes on the archwire and attached to the microimplant head with composite resin.

Listed below are the goals of skeletal open bite treatment:

- Encourage the mandible to rotate counterclockwise
- Encourage the palatal plane to rotate clockwise
- Expand the constricted maxillary dental arch
- Parallel the maxillary and mandibular occlusal planes
- Improve the upper lip–incisor relationship and smile line
- Eliminate the soft tissue parafunctions and correct speech (consult with a speech therapist)
- Reduce lower facial height to facilitate lip closure
- Achieve a normal (or deep) anterior overbite[28]

6 | Correction of Vertical Discrepancies

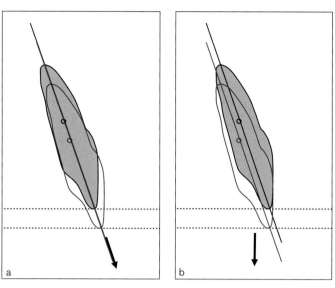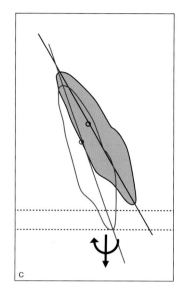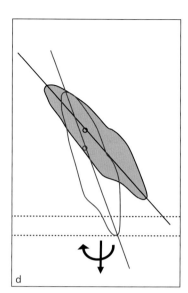

Fig 6-30 *(a to d)* Four possible methods of extruding an incisor.

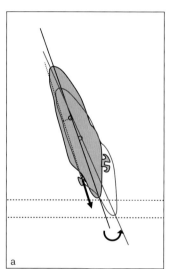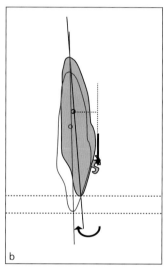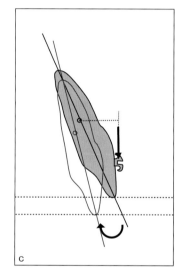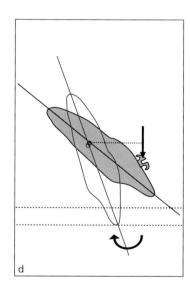

Fig 6-31 *(a to d)* The type of extrusion of an incisor depends on its axial inclination.

Extraction in the treatment of open bite

In growing patients, controlling the vertical movements of the posterior teeth is an important part of controlling the growing skeletal open bite. Eliminating premature contacts between the primary molars, or extraction, can help control a vertical growth pattern and correct the skeletal discrepancy. Mesial movement of the posterior teeth after extraction promotes counterclockwise rotation of the mandible and reduces lower facial height[28] (Fig 6-33).

Posterior bite block

A posterior bite block raising the bite over the freeway space will use muscle activity to control vertical movement of the molars, thus helping control lower facial height as well as vertical growth[29] (Fig 6-34).

Transpalatal arch and high-pull headgear combination

Vertical movement of the molars can also be controlled with the combined use of a transpalatal arch and high-pull headgear. As explained earlier, high-pull force ap-

Treatment of High-Angle Cases and Correction of Open Bite

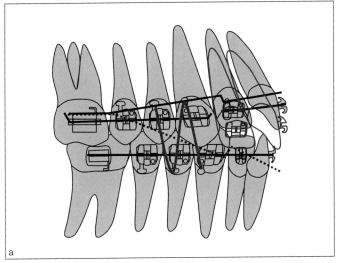
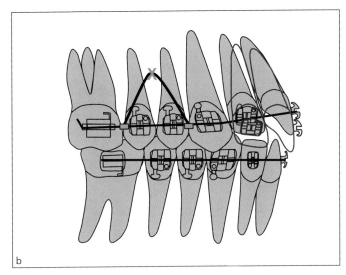

Fig 6-32 Two biomechanically efficient ways to obtain selective incisor extrusion. *(a)* A 0.017 × 0.025–inch SS segmented arch with 0.016 × 0.022–inch TMA cantilever, combined with up-and-down elastics on premolars and canines to avoid canting of the maxillary posterior segment from the counterclockwise moment. The point of application of extrusive force should be at the distal wing of the lateral incisors to obtain bodily movement. *(b)* A microimplant can be used between the maxillary premolar roots to control canting of the maxillary posterior segment. This indirect anchorage is transferred to the archwire with a V-shaped bar that is soldered onto tubes on the wire and affixed to the microimplant head with composite resin.

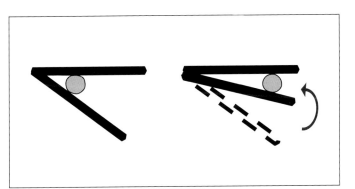

Fig 6-33 Extraction helps reduce lower facial height and encourages counterclockwise rotation of the mandible. This resembles the mechanism of a nutcracker. As the molars move mesially, the angle between the jaws decreases.[28]

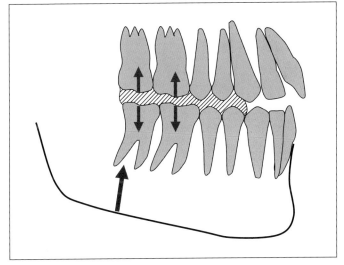

Fig 6-34 A posterior bite block can be used for vertical control of the posterior segment.

plied to the buccal aspects of maxillary molars may cause these teeth to tip buccally, the prevention of which is vital to avoid premature contacts between the palatal cusps of the maxillary molars and their antagonists, opening the bite. A transpalatal arch can be used to control molar inclinations as well as to intrude them. If the arch crosses the palatal vault 2 to 3 mm away from the mucosa, the tongue will apply vertical force on it each time the patient swallows (Fig 6-35). This function, which takes place approximately 2,400 times per day,[30] will help the molars intrude effectively with translation.

Eruption of second molars can affect vertical growth by causing premature contacts, so it is important to con-

6 | Correction of Vertical Discrepancies

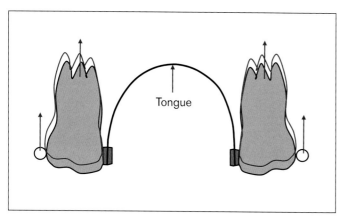

Fig 6-35 In open bite cases, control of vertical movement of the molars can be achieved effectively with a high-pull headgear–transpalatal arch combination. If the transpalatal arch crosses the palate 2 to 3 mm away from the mucosa, the molars will be intruded by vertical tongue forces during swallowing.

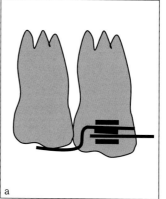 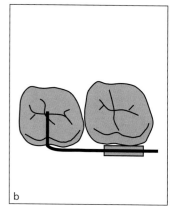

Fig 6-36 *(a and b)* In open bite cases, erupting second molars can be controlled using a 0.016 × 0.022–inch SS segmented arch that passes through the auxiliary tube of the first molar.[28]

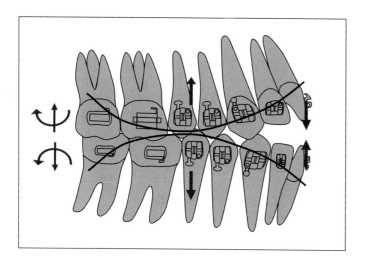

Fig 6-37 Using reverse-curved archwires to close an anterior open bite. The strong anterior box elastics prevent the premolars from erupting, while the molars intrude and tip back and the incisors extrude. These mechanics work quite effectively in a very short time, but they are heavily dependent on patient cooperation. Elastics must be worn all day, otherwise the bite may open with quick extrusion of the premolars.

trol their eruption before they reach the occlusal plane. For this purpose, a 0.016 × 0.022–inch SS segmented arch can be used as an occlusal stop[28] (Fig 6-36).

Arches with reverse curve of Spee

Anterior open bite can be closed with a combination of a reverse-curved archwire and anterior box elastics[31,32] (Fig 6-37). The archwire tends to extrude the maxillary and mandibular premolars, opening the bite, while strong anterior box elastics prevent eruption of the premolars and extrude the anteriors. Because the premolars cannot erupt, the molars intrude and tip back with reciprocal forces. These mechanics effectively close the bite in 1 or 2 months, but they are heavily dependent on patient cooperation. If the patient fails to wear the elastics, the premolars will extrude and cause the bite to open more. Even though this approach is very effective in closing the bite, the elastics should not be worn longer than 2 months because of the possibility of gingival recession and a gummy smile from overeruption of the incisors.

Molar intrusion with microimplant anchorage

Molar intrusion may be required to control the vertical discrepancy in skeletal open bite. However, using conventional techniques, this movement is one of the most challenging procedures in orthodontics—depending on strong anchorage—but intraoral anchorage is usually not

Treatment of High-Angle Cases and Correction of Open Bite

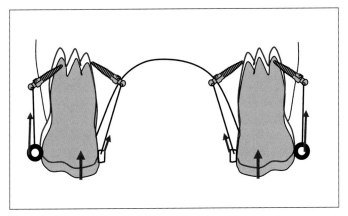

Fig 6-38 Molar intrusion using two TADs.

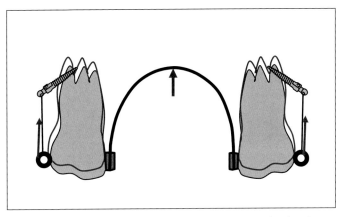

Fig 6-39 Molar intrusion with one TAD and a transpalatal arch.

Fig 6-40 Molar protraction in combination with intrusion may result in counterclockwise rotation of the mandible, thus helping to correct the skeletal open bite by reducing lower facial height.

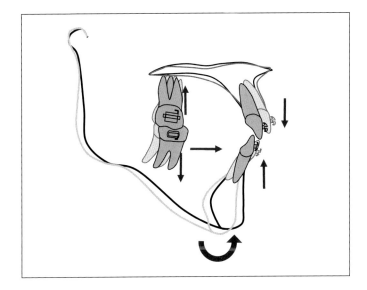

enough without extrusion of adjacent teeth. High-pull headgear with long arms in conjunction with a transpalatal arch is usually needed to achieve effective intrusion of posterior teeth (see Figs 5-7 and 6-35). Microimplant anchorage also is a very effective way to intrude molars.

There are two basic methods to intrude molars with microimplant anchorage:

- Two TADs can be inserted, both buccally and palatally, and elastic traction applied to the hooks (Fig 6-38). If two or more posterior teeth need intrusion, the force can be applied to the archwire.
- One TAD can be inserted buccally, while a transpalatal arch controls buccolingual tipping of the molar (Fig 6-39). Tongue forces during swallowing will assist this intrusion (see Fig 6-35).

In either method, TADs can be placed between the maxillary first molar and second premolar roots or between the first and second molar roots. Molar intrusion in the mandibular arch is usually more difficult than in the maxilla. Because microimplant insertion is not recommended on the lingual of the mandibular dental arch, a lingual bar can be used to control buccolingual molar inclination. Molar protraction in conjunction with intrusion causes the mandible to rotate counterclockwise and helps close the bite[28] (Fig 6-40).

A microimplant on the zygomatic cortical bone buttress is also recommended for intruding molars more effectively.[33] Even though it is stronger, zygomatic microimplant insertion requires flap surgery, which may cause soft tissue irritation.

References

1. Burstone CR. Deep overbite correction by intrusion. Am J Orthod 1977;72:1–22.
2. Burstone CJ. Modern Edgewise Mechanics and the Segmented Arch Technique. Glendora, CA: Ormco, 1995.
3. Marcotte MR. Biomechanics in Orthodontics. Toronto: Decker, 1990.
4. Levin RI. Deep bite treatment in relation to mandibular growth rotation. Eur J Orthod 1991;13:86–94.
5. Arat M, Gögen H, Parlar S, Yılmaz O, Bildir MY. Effect of Begg technique therapy in cases with excess overbite [in Turkish]. Turk J Orthod 1989;2:261–266.
6. Hans MG, Kishiyama C, Parker SH, Wolf GR, Noachtar R. Cephalometric evaluation of two treatment strategies for deep overbite correction. Angle Orthod 1994;64:265–274.
7. Schudy FF. Vertical growth versus antero-posterior growth as related to function and treatment. Angle Orthod 1964;35:75–93.
8. Perkins RA, Staley RN. Change in lip vermilion height during orthodontic treatment. Am J Orthod Dentofacial Orthop 1993;103:147–154.
9. Wylie WL. The mandibular incisor—Its role in facial aesthetics. Angle Orthod 1955;25:32–41.
10. Peck S, Peck L, Kataja M. The gingival smile line. Angle Orthod 1992;62:91–100.
11. Dong JK, Jin TH, Cho HW, Oh SJ. The esthetics of the smile. A review of some recent studies. Int J Prosthodont 1999;12:9–19.
12. Dermaut LR, Vanden Bulcke MM. Evaluation of intrusive mechanics of the type "segmented arch" on a macerated human skull using the laser reflection technique and holographic interferometry. Am J Orthod 1986;89:251–263.
13. Melsen B, Agerbaek N, Markenstam G. Intrusion of incisors in adult patients with marginal bone loss. Am J Orthod Dentofacial Orthop 1989;96:232–241.
14. Ricketts RM. Bioprogressive Therapy. Denver: Rocky Mountain Orthodontics, 1979:100.
15. Lemasson C, Labarrere H. L'ingression des incisives en technique de Root. L'orthodontie Française 1994;65:385–389.
16. Bennett JC, McLaughlin RP. Management of deep overbite with a preadjusted appliance system. J Clin Orthod 1990;24:684–696.
17. Parker CD, Nanda RS, Currier GF. Skeletal and dental changes associated with the treatment of deep bite malocclusion. Am J Orthod Dentofacial Orthop 1995;107:382–393.
18. Eberhart BB, Kuftinec MM, Baker IM. The relationship between bite depth and incisor angular change. Angle Orthod 1990;60:55–58.
19. Dermaut LR, De Pauw G. Biomechanical aspects of Class II mechanics with special emphasis on deep bite correction as a part of the treatment goal. In: Nanda R (ed). Biomechanics in Clinical Orthodontics. Philadelphia: Saunders, 1996:90.
20. Ferguson JW. Lower incisor torque: The effects of rectangular archwires with a reverse curve of Spee. Br J Orthod 1990;17:311–315.
21. Lee JS, Kim JK, Park YC, Vanarsdall RL Jr. Applications of Orthodontic Mini-implants. Chicago: Quintessence, 2007:156.
22. Burstone CJ. How to level the occlusal plane in deep bite cases. Presented at the 4th International Orthodontic Congress, San Francisco, 12–17 May 1995.
23. Burstone CJ. Contemporary management of Class II malocclusions: Fact and fiction in Class II correction. In: Nanda R (ed). Biomechanics in Clinical Orthodontics. Philadelphia: Saunders, 1997:246–256.
24. Mulligan TF. Common Sense Mechanics in Everyday Orthodontics. Phoenix: CSM, 1998.
25. Philippe J. Orthodontie, des principes et une technique. Paris: Prélat, 1972.
26. Tuncer AV, Işıksal E, Tosun Y. Cephalometric study of the tongue posture in different anomalies [in Turkish]. Turk J Orthod 1992;5:113–120.
27. Tosun Y, Tuncer AV, Tosun Ş. Skeletal dimension of anterior open bite [in Turkish]. Turk J Orthod 1991;4:46–51.
28. Tosun Y. Biomechanical Principles of Fixed Orthodontic Appliances. İzmir, Turkey: Aegean Univ, 1999:195–197.
29. Koralp E, İşçan H. Effect of the application of passive posterior bite block in combination with vertical chin cap in the treatment of open bite on the facial vertical dimensions and dento-alveolar structures [in Turkish]. Turk J Orthod 1991;4:55–61.
30. Graber TM. Orthodontics: Principles and Practice, ed 3. Philadelphia: Saunders, 1972:169.
31. Enacar A, Uğur TA, Toroglu S. A method for correction of open bite. J Clin Orthod 1996;30:43–48.
32. Aras A, Çinsar A. Treatment of an adult anterior open bite using elastics and rectangular NiTi wires in Spee arch form [in Turkish]. Zeitschrift der Türkischen Zahnarzte 1997;2:41–42.
33. Erverdi N, Tosun T, Keles A. A new anchorage site for the treatment of anterior open bite: Zygomatic anchorage case report. World J Orthod 2002;3:147–153.

Correction of Transverse Discrepancies

CHAPTER 7

This chapter discusses the second of the three major discrepancies (vertical, transverse, and anteroposterior) that might be encountered in planning treatment for the orthodontic patient. Two of the most common transverse discrepancies are anterior and posterior crossbites. Skeletal and dental asymmetries are also important issues to consider in the overall treatment plan.[1]

Crossbite Correction

At the first stage of diagnosis and treatment planning, dental or skeletal crossbites should be considered along with the orthopedic problem because they can affect mandibular movements, temporomandibular joint (TMJ) function, and periodontal health such as gum recession and loss of bony support owing to traumatic occlusion (Fig 7-1). Posterior crossbites are mainly the result of maxillary dental arch constriction, either unilaterally or bilaterally; thus, expansion of the maxillary arch is necessary to obtain a harmonious transverse relationship within both dental and skeletal structures. If the skeletal posterior crossbite is in question, the expansion should be done with a fixed expander.[1]

Rapid maxillary expansion

The rapid maxillary expander (RME) has been used for years to obtain skeletal expansion by opening the midpalatal suture to widen the transverse dimensions of the maxillary arch. The most common and hygienic type of expander is the hyrax screw. In the transitional dentition, only the maxillary first molars are banded, and the arms of the appliance are extended up to the primary canines (Fig 7-2).

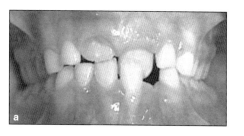
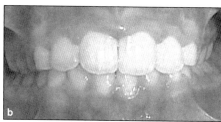

Fig 7-1 (a) Traumatic occlusion due to anterior crossbite in the transitional dentition. (b) The problem was corrected easily using a removable appliance.

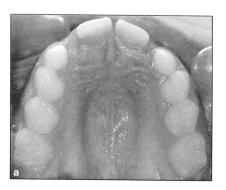
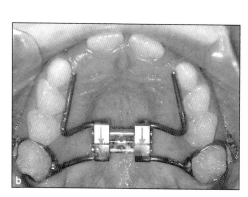

Fig 7-2 In the transitional dentition (a), the arms of the hyrax screw RME can be extended up to the canines (b).

In children and adolescents, the screw is turned a quarter turn once or twice a day; theoretically, an expansion of 0.25 to 0.5 mm/day can be obtained. In each quarter turn, approximately 0.9 to 4.5 kg of force is applied to the teeth.[2] The expansion force affects all the facial sutures, especially the midpalatal suture. As expansion progresses, a large median diastema appears, indicating opening of the suture. This diastema closes spontaneously with the pull of the transseptal fibers. As the transverse dimension of the maxillary dental arch increases, the palatal vault depth decreases, increasing the volume of the nasopharyngeal airway.[1,3] At the end of the expansion process, some dental and skeletal relapse is expected as a result of buccal tipping of the teeth as well as resistance of the surrounding musculature. Overexpansion is usually necessary until the palatal cusps of the maxillary teeth contact the lingual aspects of the buccal cusps of the mandibular molars to compensate for the relapse. The expander should be kept in place for at least two-thirds of the overall duration of treatment (ie, approximately 9 to 10 months) for stability.

When the amount of force is over the optimum limit necessary for tooth movement, skeletal expansion is observed before tooth movement. The most suitable time for transverse expansion is either the primary or transitional dentition.[4–12] Most authors agree that rapid maxillary expansion should be performed before the midpalatal suture closes. The time of ossification of the midpalatal suture, however, is controversial.[13] Rapid expansion can also be applied at later ages; however, bone density in adults limits the stability as well as the range of transverse expansion.[6,12,14–18]

During rapid maxillary expansion, the maxillary alveolar bones rotate around a point in the nasal cavity near the frontomaxillary suture, creating a triangle with the apex oriented superiorly[19] (Fig 7-3). In addition, it has been shown that A-point comes forward as a result of expansion.[20,21] As the molars and premolars tip buccally, the palatal cusps move downward and make premature contact with their antagonists, stimulating clockwise rotation of the mandible. In high-angle cases, lower facial height may increase and the soft tissue profile may worsen. It has been shown that 30% of patients presenting with clockwise rotation after rapid maxillary expansion return to their original condition; however, 30% remain the same, and 40% continue to open.[22] Bite opening due to expansion can be avoided with the use of an expander combined with a posterior bite block.[23] Correcting buccal tipping of the posterior teeth by placing buccal torque in the archwire or transpalatal arch

Fig 7-3 During rapid maxillary expansion, the maxillary alveolar bones rotate around a point in the nasal cavity near the frontomaxillary suture, creating a triangle with the apex oriented superiorly. (Redrawn from Majourau and Nanda[19] with permission.)

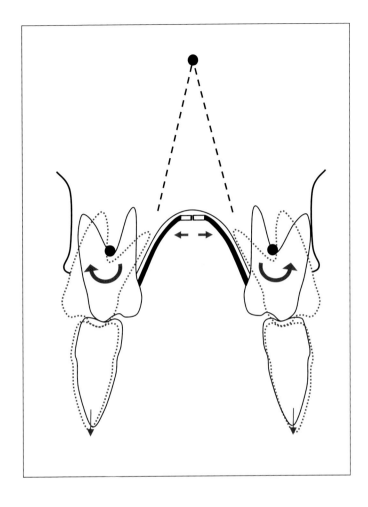

results in extrusion of the buccal cusps rather than intrusion of the palatal cusps, so it does not eliminate the increase in lower facial height. That can be done with posterior bite blocks, which control facial height by limiting extrusion of the palatal cusps.

Transpalatal arch

Posterior crossbite can also be corrected with a removable transpalatal arch. When the U-loop in the middle is activated (ie, opened) and the arch is engaged in the palatal tubes on the molar bands, it applies buccal force on the molars, causing buccal tipping. Further activation of the transpalatal arch can be done intraorally with three-jaw pliers (Fig 7-4). After expansion, buccal root torque should be placed in the arch to correct molar inclinations. Note that some extrusion of the palatal cusps as a result of expansion should be expected.

Quad helix

The quad helix can be made from a 0.036-inch wire. Four helices give the arch increased flexibility and working range. The quad helix is activated extraorally before placement. Its main effect is on the molars; however, the lateral arms can also be activated to expand premolars and canines. If further activation is necessary, it can be performed intraorally by means of three-jaw pliers. Figure 7-5 shows the intraoral activation and adjustment of the arch. A V-bend in the anterior bridge causes the molars to rotate mesiopalatally and tip buccally.[1] This rotation can be compensated for by subsequent bends in the lateral bridges. To avoid balancing forces and obtain equal and opposite moments, an equal amount of activation (ie, of bend given on the wire) should be applied on each side.

7 | Correction of Transverse Discrepancies

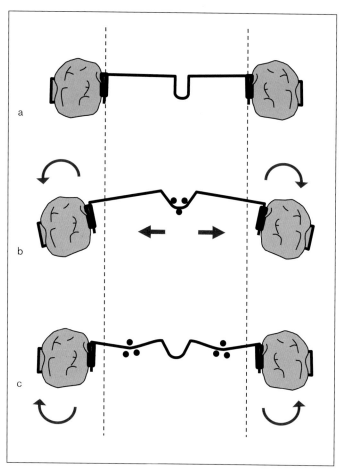

Fig 7-4 (a) Transpalatal arch for transverse expansion of the molars. (b) When the U-loop is activated, the molars rotate distobuccally and tip buccally. (c) To eliminate the rotation, compensating bends can be made intraorally using three-jaw pliers.

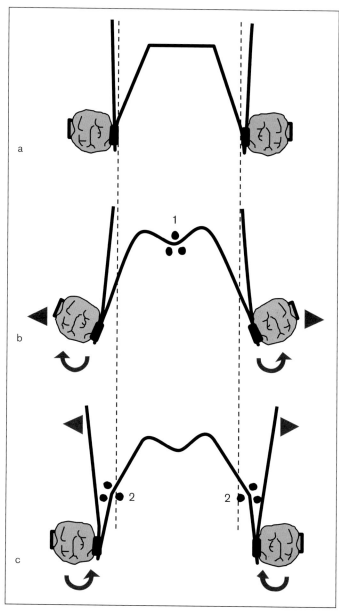

Fig 7-5 (a) Quad helix in place. (b) Initial intraoral activation consisting of placing a V-bend (1) in the anterior bridge, which causes the molars to rotate mesiopalatally and tip buccally. (c) To compensate for the rotation, adjustment bends are made on the lateral bridges (2).

The quad helix can be used in the transitional as well as the permanent dentition. In the early transitional dentition, sutural opening can be obtained in addition to dental expansion. In the permanent dentition, the main effect of the appliance is buccal tipping, though some corrective buccal root torque can be applied to the molars. Therefore, the initial inclinations of the posterior teeth need to be considered in appliance selection. If the molars are already tipped buccally, then rapid maxillary expansion is the treatment of choice to avoid further tipping and subsequent periodontal consequences such as recession.

Fig 7-6 Crossbite elastics can be used to correct posterior crossbite. *(a)* The extrusive effect of the elastic must be considered when treating high-angle patients. *(b)* Microimplant anchorage can be used to avoid the extrusive effect of the crossbite elastic.

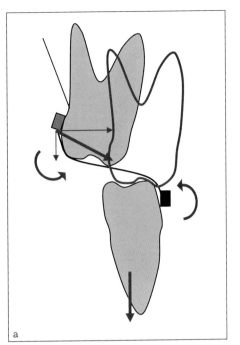
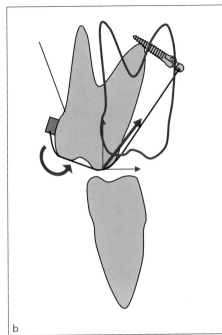

Elastic archwires

It is well recognized that elastic archwires tend to widen the dental arch, especially straight, multistrand wires such as twist-flex wires, which expand the molar-to-molar dimension and help correct an edge-to-edge bite or a posterior crossbite. This is a desired effect in nonextraction cases; however, in the long term, flexible archwires may not only expand the posterior segment but also the canine-to-canine dimension. In a patient 7 to 8 years of age, in whom the incisors are erupting, the mandibular intercanine dimension expands naturally, then remains stable throughout life. Any attempt to increase this dimension tends to relapse.[24–26] Thus it is important to maintain the original transverse dimension between canines throughout appliance treatment for stability concerns.

Crossbite elastics

The crossbite elastic is a convenient way to correct posterior crossbites. The elastic is attached between the palatal aspect of the maxillary molar and the buccal aspect of the mandibular molar. Because this mechanism works with reciprocal anchorage, the mandibular molar anchorage should be reinforced with a lingual arch if constriction of the mandibular posterior segment needs to be avoided. Fig 7-6a shows how the resultant force generated by elastics tends to extrude the molars as they tip.[1] Posterior crossbites can be effectively corrected by using microimplant anchorage, which negates the elastics' extrusive effect (Fig 7-6b).

Scissors Bite Correction

Scissors bite, wherein the palatal surface of a maxillary molar or premolar is in contact with the buccal surface of a mandibular molar or premolar, can involve one or several teeth. A careful diagnosis involving the vertical growth pattern of the patient is imperative for treatment planning. Usually scissors bite consisting of one tooth can be corrected easily with straight-wire mechanics using flexible archwires. Crossbite elastics help boost the effect of the wire; however, one should remember the adverse effects related to the vertical force component of

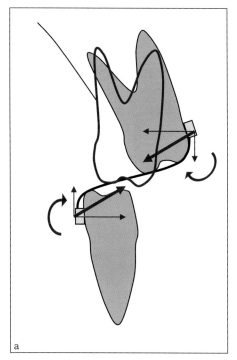 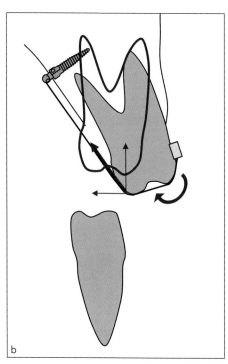 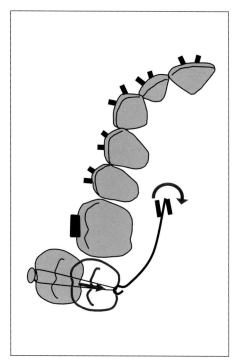

Fig 7-7 *(a)* Crossbite elastics used to correct scissors bite cause molar extrusion. *(b)* Microimplant anchorage helps counteract the extrusive effect of the crossbite elastics while helping to correct the scissors bite.

Fig 7-8 A bracket-head screw can be used in conjunction with a cantilever to apply force in the desired direction.

these elastics (Fig 7-7). If the second molar is in scissors bite, a 0.016-inch nickel titanium (NiTi) or titanium-molybdenum alloy (TMA) cantilever combined with a full-size stainless steel (SS) archwire works reasonably well. If the scissors bite involves several teeth, it may be due to an overexpanded or deformed maxillary arch or a constricted mandibular arch. The expanded maxillary buccal segment can be corrected with both an archwire-cantilever combination and crossbite elastics. Mandibular arch expansion is usually difficult to accomplish because of the powerful musculature and high-density bone surrounding the teeth. A 0.032-inch lingual arch can expand the mandibular molar-to-molar dimension with tipping and extrusion. In high-angle cases, a posterior bite block or microimplant should be used to control vertical movement of the molars. Sung et al[27] suggest using a bracket-head screw to apply force in the proper direction. For either side of the dental arch, right- or left-handed screws can be used to make sure that the moment applied to the screw moves it deeper into the bone (Fig 7-8).

References

1. Tosun Y. Biomechanical Principles of Fixed Orthodontic Appliances. Izmir, Turkey: Aegean University, 1999:164–165,211, 214–216.
2. Isaacson RJ, Wood JL, Ingram AH. Forces produced by rapid maxillary expansion. Angle Orthod 1964;34:256–270.
3. Tosun Y. Effect of rapid expansion on palatal dimensions. J Aegean Univ 1991;12:97–100.
4. da Silva Filho OG, Magro AC, Capelozza Filho L. Early treatment of the Class III malocclusion with rapid maxillary expansion and maxillary protraction. Am J Orthod Dentofacial Orthop 1998;113:196–203.
5. Biederman WB, Chem B. Rapid correction of Class 3 malocclusion by midpalatal expansion. Am J Orthod 1973;63:47–55.
6. Bishara SE, Staley RN. Maxillary expansion: Clinical implications. Am J Orthod Dentofacial Orthop 1987;91:3–14.
7. Brin I, Hirshfeld Z, Shanfeld JL, Davidovitch Z. Rapid palatal expansion in cats: Effects of age on sutural cyclic nucleotides. Am J Orthod 1981;79:162–175.
8. Cleal JF, Bayne DI, Posen JM, Subtelny JD. Expansion of the midpalatal suture in the monkey. Angle Orthod 1965;35:23–35.
9. Cotton LA. Slow maxillary expansion: Skeletal versus dental response to low magnitude force in *Macaca mulatta*. Am J Orthod 1978;73:1–23.
10. Storey E. Tissue response to the movement of bones. Am J Orthod 1973;64:229–247.
11. Ten Cate AR, Freeman E, Dickinson JB. Sutural development: Structure and its response to rapid expansion. Am J Orthod 1977;71:622–636.
12. Wertz RA. Skeletal and dental changes accompanying midpalatal suture opening. Am J Orthod 1970;58:41–66.
13. Persson M, Thilander B. Palatal suture closure in man from 15 to 35 years of age. Am J Orthod 1977;72:42–52.
14. Enacar A, Demirhano lu M, Özgen M. Rapid maxillary expansion in adults. Turk Ortodonti Derg 1993;6:64–71.
15. Tosun Y, Tuncer AV. Rapid palatal expansion in an adult Class III case (case report). Turk Ortodonti Derg 1991;4:89–94.
16. Bell RA. A review of maxillary expansion in relation to rate of expansion and patient's age. Am J Orthod 1982;81:32–37.
17. Isaacson RJ, Ingram TD. Some effects of rapid maxillary expansion in cleft lip and palate patients. Angle Orthod 1964;34:143–154.
18. Zimring JF, Isaacson RJ. Forces produced by rapid maxillary expansion. III. Forces present during retention. Angle Orthod 1965;35:178–186.
19. Majourau A, Nanda R. Biomechanical basis of vertical dimensional control during rapid palatal expansion therapy. Am J Orthod Dentofacial Orthop 1994;106:322–328.
20. Haas AJ. Long-term posttreatment evaluation of rapid palatal expansion. Angle Orthod 1980;50:189–217.
21. Davis WM, Kronman JH. Anatomical changes induced by splitting of the midpalatal suture. Angle Orthod 1969;39:126–132.
22. Wertz R, Dreskin M. Midpalatal suture opening: A normative study. Am J Orthod 1977;71:367–381.
23. Aras A, Sürücü R. Comparative study of the expansion effectiveness of modified Haas appliance with bite block and rapid expansion. Turk Ortodonti Derg 1990;3:14–20.
24. Moorrees CFA, Gron AM, Lebret LM, Yen PK, Fröhlich FJ. Growth studies in the dentition: A review. Am J Orthod 1969;55:600–616.
25. Bishara SE, Jacobsen JR, Treder J, Nowak A. Arch width changes from 6 weeks to 45 years of age. Am J Orthod Dentofacial Orthop 1997;111:401–409.
26. Moorees CFA, Chadha JM. Available space for the incisors during dental development–A growth study based on physiologic age. Angle Orthod 1965;35:12–22.
27. Sung JH, Kyung HM, Bae SM, Park HS, Kwon OW, McNamara JA Jr. Microimplants in Orthodontics. Daegu, Korea: Dentos, 2006.

Correction of Anteroposterior Discrepancies

CHAPTER 8

The Class I molar relationship is one of the most important goals for the clinician seeking to achieve a functional and balanced occlusion.[1] A Class II molar relationship can be corrected by distal movement of the maxillary molars, mesial movement of the mandibular molars, or a combination of both.

Molar Distalization

In some nonextraction treatments, molar distalization, or retraction, is the method of choice to gain space in the dental arch. The decision to distalize should be based on a careful examination and treatment plan. Several criteria should be considered before distally moving molars:

- Axial inclination of the molars
- Availability of space for the second and third molars
- Rotation of the molars
- Patient's vertical growth pattern

Molar distalization is usually performed to gain 2 to 3 mm of space in the dental arch to obtain a Class I relationship. Because the distalizing force is usually applied to the buccal tube away from the center of resistance, it is easier to retract mesially tipped and mesiolingually rotated molars to where they are supposed to be than it is to move them bodily. Excessive retraction, however, may jeopardize the second and third molars—possibly impacting them.[2,3]

Molar distalization can be effectively accomplished in the transitional dentition before the second molars fully erupt and when the alveolar bone is relatively active.[4] This is the best time to get good bone response to orthodontic treatment. If the third molars are missing, the erupting second molars that are pushed distally by the first molar can upright naturally and subsequently accommodate with a normal inclination. If the second molars are fully erupted, they will be tipped distally to an unstable inclination, causing marginal ridge discrepancies[5] as well as relapse after retraction is completed. It is therefore necessary to consider including the second molars in the appliance and retracting them before or with the first molars.

Regardless of the distalizing method, the patient's vertical growth pattern is an important point to consider. During molar distalization, some bite opening can be expected because of premature contacts. In normal- to low-angle patients, this is desired as the mandible rotates and corrects the profile by increasing lower facial height. In high-angle patients, however, clockwise rotation may worsen the profile and cause bite opening. On the other hand, bruxism and clench-

8 | Correction of Anteroposterior Discrepancies

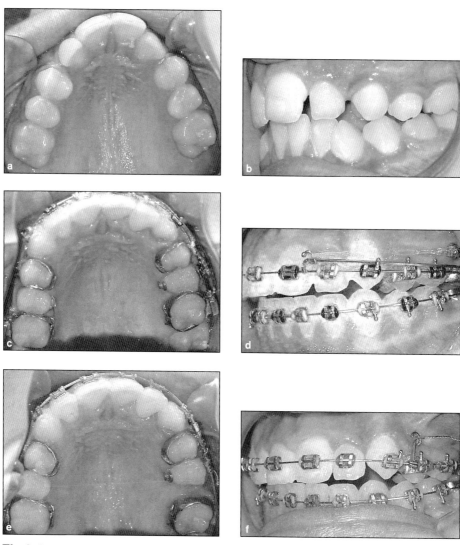

Fig 8-1 Microimplant anchorage mechanics for molar distalization. *(a and b)* Before treatment. *(c and d)* Molar distalization with chain elastic and sliding jig. *(e and f)* Effective molar distalization after 4 months.

ing, like excessive occlusal forces, may act to reduce—or even prevent—distalization.[5]

Molar distalization mechanisms include extraoral appliances, microimplants (temporary anchorage devices [TADs]), 2 × 4 arches, Nance appliance and coil spring combinations, superelastic wires, and sliding jigs.

Extraoral appliances

In normal- to low-angle patients, cervical or combined headgear can be used to distalize the molars effectively. These types of headgear need to be worn approximately 16 hours a day, requiring excellent patient cooperation. If the goal is to gain space on the arch by individual molar movement only, 300 to 350 g may be sufficient. When the force is higher than the optimum level needed for individual tooth movement, the molars can be used as a "handle" to obtain an orthopedic effect on the maxillary bone.[6] Orthopedic as well as dental changes can be obtained with the use of 400 to 600 g of force for an average of 16 hours per day in the transitional dentition.[7]

Extraoral force is applied to the molar by means of a facebow. In the transitional dentition, as the molars move back, the primary teeth are also pulled back by means of the transseptal fibers. When the permanent

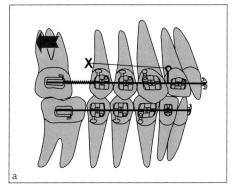 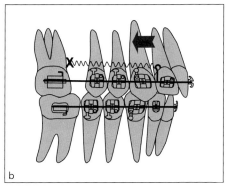 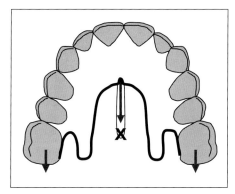

Fig 8-2 *(a)* Molar distalization with open coil spring and microimplant anchorage. *(b)* After molar distalization, the TAD needs to be removed and replaced mesial to the molar roots to retract the anterior teeth.

Fig 8-3 Molars can be distalized by taking support from the TAD inserted into the midpalatal suture.

teeth come through, they erupt into a Class I relationship. Sometimes the effect of the transseptal fibers reaches the anterior segment and effects spontaneous correction of mild incisor crowding.[7] In normal to severe anterior crowding cases, however, extraoral force combined with 2 × 4 arch mechanics would improve the result. Molar distalization with extraoral force might provide enough room to rule out a second-phase comprehensive treatment in the permanent dentition.

Theoretically, if the line of action of the extraoral force passes through the center of resistance of the molar, it will cause bodily movement. In practice, however, bodily movement is difficult to achieve because the headgear is a sort of removable appliance; thus the line of action of the force may need adjusting every time it is worn to keep it constant. Usually a few tipping and uprighting movements are observed during the retraction process. If the crown is tipped excessively backward, the outer arms need to be adjusted so the force passes above the center of resistance to correct the roots. Three-dimensional control of the first and second molars is imperative to achieve functionally stable occlusion in the posterior segment.

Microimplants

In cases where second molars are fully erupted, molar distalization is a difficult maneuver requiring strong anchorage in the anterior segment. Microimplants provide the absolute anchorage needed to accomplish effective distal movement of the maxillary dentition along with the molars. A TAD is usually placed between the first molar and second premolar roots from which distal traction can be applied on a sliding jig (Fig 8-1) or a sliding hook attached to an open coil spring pushing the molar (Fig 8-2a). The direction of the force passes close to the center of resistance; thus, minor molar tipping and distopalatal rotation may occur. After distalization has been completed, if the first TADs interfere with the second premolar roots, they can be removed and a second set of TADs can be placed just mesial to the first molar roots to make room for retraction of the anterior teeth (Fig 8-2b).

Molar retraction can also be accomplished with a transpalatal arch that takes support from the microimplant inserted in the midpalatal area (Fig 8-3). This area provides excellent cortical bone support for primary stability. In young patients whose suture is still open, the microimplant can be placed adjacent to the suture. The advantage of this area is that it provides abundant space such that the force can be adjusted according to the type of movement desired.[5] In shallow palatal vaults, the line of action passes through the center of resistance. Palatal TADs cause more discomfort than the buccal ones because they interfere with the tongue.

2 × 4 arches

The tipback effect generated by 2 × 4 arches can be used to gain 1 to 2 mm of space in the arch of the transitional dentition (see Figs 6-26 and 6-27). This is usually enough to provide room for erupting permanent teeth; however, the distal inclination of the molars is not stable, and they tend to relapse when the second molars

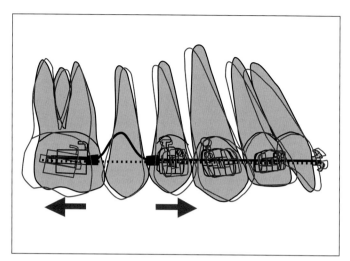

Fig 8-4 Use of superelastic archwires in maxillary molar distalization. The amount of activation of this arch is about the length of a molar tube (approximately 6 mm) (heavy line). As the wire straightens, it applies a force of approximately 100 g to both sides. Class II elastics (100 to 150 g) are used to prevent protrusion of the maxillary anteriors. (Redrawn with permission from Locatelli et al.[9])

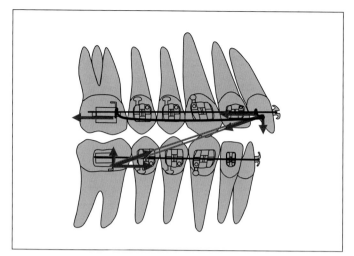

Fig 8-5 Using a sliding jig for maxillary molar distalization. The horizontal component of the Class II elastic force is transferred to the molar by means of the jig.

erupt. In such cases, the first molar roots must be corrected later with headgear.

Nance appliance and coil springs

Nickel titanium (NiTi) coil springs in combination with a Nance appliance on the first premolars is a convenient way to distalize molars. However, because the palate is not a reliable anchorage site, it may not resist the distally directed forces; thus, some incisor protrusion can be expected as a result of molar movement.[8]

Particularly shallow palatal vaults cannot provide sufficient anchorage support because the Nance acrylic button tends to slide over the palatal mucosa. Uprighting springs combined with vertical slotted brackets and heavy Class II elastics on the canines can be used to reinforce premolar anchorage (see Fig 5-2). Gianelly et al[8] do not recommend anchorage reinforcement if the incisor protrusion is less than 2 mm. If it is more than 2 mm, they suggest using 100-g Class II elastics and putting 10 to 15 degrees of labial root torque on the mandibular incisors to support mandibular anchorage. They also suggest moving the molars into a Class III relationship to compensate for subsequent anchorage loss.

After distalization, headgear with long arms should be used to upright distally tipped molars and prevent relapse. Retraction of the premolars should be started immediately with a sliding jig–Class II elastics combination along with headgear application.

Superelastic wires

Locatelli et al[9] suggest using superelastic NiTi archwires to distalize the molars (Fig 8-4). A 0.018 × 0.025–inch NiTi archwire deflected between the first premolar and first molar applies 100 g to each side. Using 100- to 150-g Class II elastics will prevent protrusion of the anterior teeth. With this method, the molars should move about 1 mm per month.

Sliding jig

The sliding jig is bent from 0.7-mm stainless steel (SS) wire. Combined with Class II elastics, the sliding jig is used to distalize molars in the permanent as well as transitional dentition (Fig 8-5). As the molars move distally, the premolars are pulled distally by means of the transseptal fibers. The horizontal component of the elastic force can be increased if the hook of the jig is placed mesial to the lateral incisor. The jig is deflected by the distal force, which tends to widen the anterior segment of the arch. This widening effect is negligible if contin-

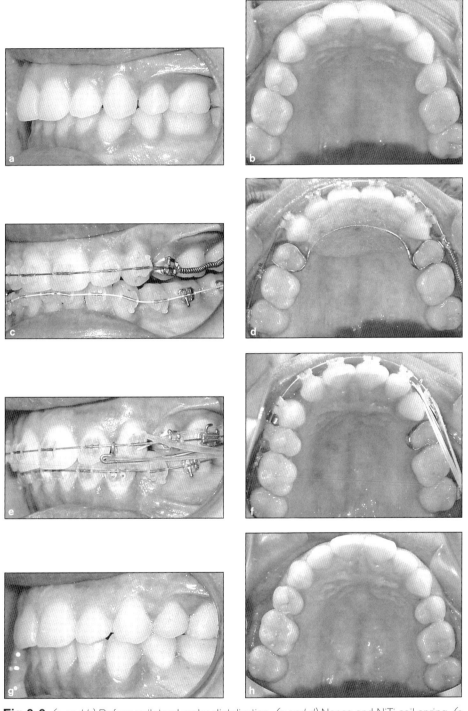

Fig 8-6 *(a and b)* Before unilateral molar distalization. *(c and d)* Nance and NiTi coil spring. *(e and f)* Sliding jig, chain, and Class II elastics combination. *(g and h)* After treatment.

uous 0.016 × 0.022–inch SS wire is used; the effect can be further resolved with subsequent rectangular wires.

One should also be careful about the long-term side effects of Class II elastics. To avoid extrusion of mandibular molars and the maxillary anterior segment, they should be used on stiff rectangular archwires. The sliding jig can be kept in place after molar distalization to move the premolars distally one by one, combined with buccal and palatal chain elastics (Fig 8-6).

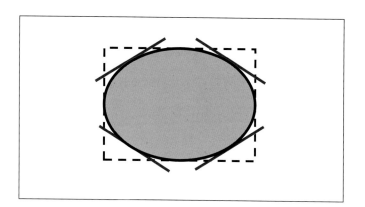

Fig 8-7 Rounding the ends of rectangular SS wires used for en masse retraction permits the wire to slide easily through the tubes. Rounding the corners of the wire prevents loss of control from overreduction of its cross section.

Molar Protraction

Protraction of molars is easier in the maxilla than in the mandible owing to the relative abundance of trabecular bone in the former. The large amount of cortical bone and the surrounding powerful musculature in the posterior mandible adversely affect anteroposterior molar movement, which becomes even more difficult with time as the alveolar bone narrows.

Molar protraction to an edentulous area is especially difficult in the mandible.[5] It is also more difficult in adults than in children. Children and young adults have fewer periodontal and root resorption problems during space closure than do older adults.[10] These factors must be considered in treatment planning as well as in mechanical application. Because of biologic limitations and consequent mesial tipping, protraction of mandibular molars is a complex procedure requiring patience and careful control.

If the molars are to be moved along a continuous archwire, a 0.016 × 0.022–inch or 0.017 × 0.025–inch SS wire (in 0.018-inch slots) is recommended to avoid mesial tipping and undesired retraction of the anterior teeth. Lingual root torque in the incisors will reinforce the anchorage. A NiTi push-coil spring between the first and second molar is recommended to avoid undesired anterior movement and to maintain overjet and overbite. Anchorage needs to be reinforced by 3/16- or 1/4-inch, medium Class II elastics between the mandibular second molars and maxillary canines (or lateral incisors).

To reduce the friction between wire and molar tube, the ends of the wire should be rounded with a diamond bur and polished with a rubber wheel before insertion (Fig 8-7).[11] If frictionless mechanics are preferred, a 0.017 × 0.025–inch SS wire with a closing loop can be used. A 5-degree distal tipback and toe-in should be made to avoid mesial tipping and mesiolingual molar rotation. It is important to maintain the axial inclination of the molar by controlling the amount of activation of the loop. A Bull loop with 1 mm of activation per month in children and adolescents and the same amount of activation every 2 months in adults is fail-safe.

The patient's vertical growth pattern also needs to be considered in treatment planning and mechanical application. During protraction, some premature contacts in the posterior segment or molar extrusion (because of mesial tipping) may affect the vertical dimensions of the face. In normal- to low-angle patients, molar extrusion is helpful to open the bite. In high-angle or skeletal openbite patients, however, molar extrusion needs to be controlled by careful mechanical application. In these cases, molar intrusion before or during protraction should be accomplished to avoid bite opening and associated side effects. This can be effectively done with a posterior bite block or microimplant anchorage mechanics.

Microimplants offer good intraoral anchorage in molar protraction mechanics. They are usually inserted between the canine and first premolar roots. The force is applied directly to the molars' power hooks, so the line of action passes through the center of resistance of the teeth. By this means, the molars move mesially by sliding along the rectangular archwire (preferably 0.018 × 0.025–inch SS wire in an 0.018-inch slot), rounded and polished posteriorly to ease sliding (Fig 8-8a). In addition, a light force can be applied lingually to prevent rotations (Fig 8-8b).

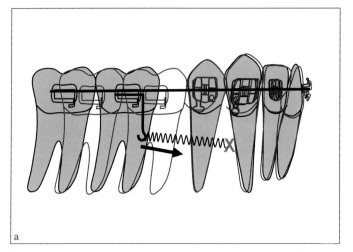
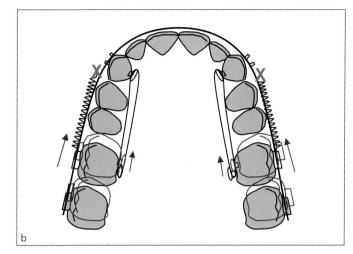

Fig 8-8 Molar protraction using microimplant anchorage. *(a)* Coil spring force can be applied between the TAD inserted anteriorly and the power hook on the molar to obtain translation. *(b)* In addition to buccal force, a light force should be applied on the lingual aspect to prevent molar rotation.

Uprighting Mesially Tipped Molars

If the mesial contact breaks, molars tend to tip mesially. This is especially true for the mandibular molars, which are already slightly mesially inclined. Tipped molars can cause several functional and periodontal problems, such as temporomandibular joint dysfunction, undesired bite opening, and alveolar bone loss owing to premature contacts. Molar uprighting is therefore important to avoid these adverse effects, to establish a more balanced occlusion, and to provide suitable relationships with the adjacent teeth for subsequent prosthetic options.

Molar uprighting mechanics depend on the treatment goals. If the space mesial to the molar is to be opened for future prosthetic restoration, uprighting can be accomplished with a flexible straight wire passing through the tube (Fig 8-9a). With the counterclockwise moment generated by the straight wire, the crown tips backward and extrudes. A looped archwire or cantilever spring also has an extrusive effect on the tooth (Figs 8-9b and 8-9c). Bending a step mesial to the molar to compensate for the extrusion will create a clockwise moment that tends to tip the molar farther mesially (Fig 8-10).

Molar extrusion is usually not desired in high-angle patients, as it may cause bite opening. In this case, uprighting should be performed in conjunction with intrusion. In conventional mechanics, molar uprighting with intrusion is quite complicated. If the tipping angle of the molar relative to the occlusal plane is equal to the anterior angle, the moments are equal; thus, there is no balancing force on the system (Fig 8-11a). It is important that the posterior moment be less than the anterior moment to obtain an intrusive force on the posterior segment. To do so, the anterior angle (x) should be higher than the tipping angle of the molar (y) (Fig 8-11b).

If the space mesial to the molar is to be closed, distal tipping of the crown should be prevented with a laceback or cinchback. In this case, the counterclockwise moment tends to move the roots mesially around the center of rotation located at the crown (Fig 8-12). Protraction of the molar can be started after uprighting is completed. A stiff straight wire can then be engaged in the molar tube and a mesial force applied by means of a laceback or coil spring. A prefabricated uprighting spring has been shown to be effective in space closure and bite control in skeletal open bite cases.[12]

To upright molars with intrusion, various segmented wires such as 0.016 × 0.022–inch SS with a helix or 0.017 × 0.025–inch titanium-molybdenum alloy (TMA) can be used. The anterior segment should be incorporated with a full-size SS wire to reinforce anterior

8 | Correction of Anteroposterior Discrepancies

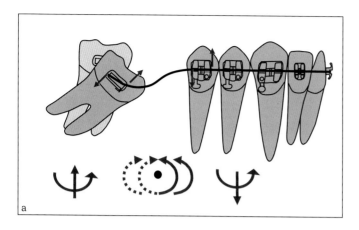

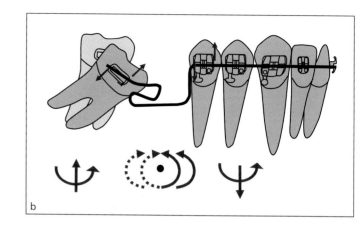

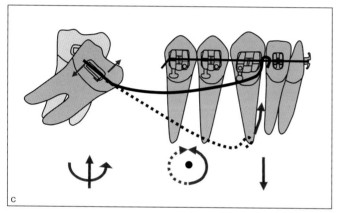

Fig 8-9 If a flexible straight wire is engaged in the mesially tipped molar (a), the tooth uprights with extrusion. A looped wire (b) or a cantilever spring (c) has the same extrusive effect on the molar.

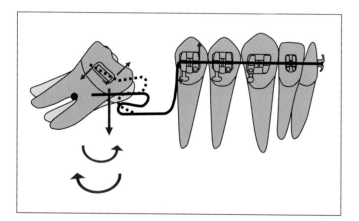

Fig 8-10 When uprighting molars with a looped archwire, some clinicians make incorrect steps in the loop to intrude the molar. This bend causes further mesial tipping because of the downward force passing mesial to the molar's center of resistance.

anchorage. To avoid extrusion and clockwise rotation of the anterior segment and apply optimum intrusive force on the molar, the difference between moments should be as low as possible. This adverse effect can be prevented by light to medium, 5/16-inch anterior box elastics. Labial root torque can also prevent incisor protrusion.

In severe skeletal open bite cases, a posterior bite block may be needed to control molar extrusion during uprighting when conventional mechanics are not sufficient (see Fig 6-34). It is important that the bite block be used 12 to 16 hours per day or longer to get an effective response.

Uprighting Mesially Tipped Molars

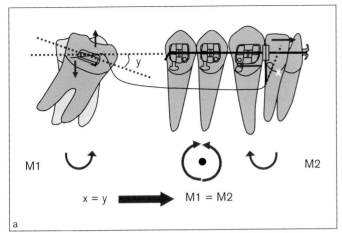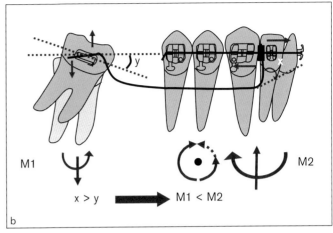

Fig 8-11 In high-angle cases, molar extrusion needs to be avoided during uprighting. *(a)* If the x and y angles are equal, then the anterior and posterior moments are equal (M1 = M2), thus no balancing forces occur. *(b)* To obtain intrusive force on the molar, the anterior angle (x) needs to be increased. The anterior segment should be incorporated with a full-size SS wire to reinforce anterior anchorage. The amount of anterior moment must be controlled essentially with the angulation of the wire to avoid adverse effects at the anterior segment. If this moment is too high, it could result in extrusion of the premolars and protrusion of the mandibular incisors. This adverse effect can be controlled be reducing the angle or with the use of light or medium, $\frac{5}{16}$-inch anterior box elastics. In addition, labial root torque may prevent incisor protrusion.

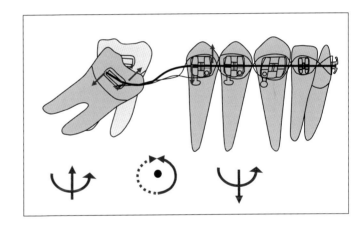

Fig 8-12 If the space mesial to a molar is to be closed, distal crown movement should be prevented with a laceback or cinchback to obtain mesial root movement. Protraction of the molar should be done after uprighting is completed.

Microimplant anchorage mechanics for molar uprighting

Molar uprighting is a difficult movement that needs consistent mechanics, strong anchorage preparation, and patience. To avoid extrusion, leveling and alignment of the anterior teeth with up to full-size SS wire are needed before applying the uprighting moment (see Fig 8-11b). Molar uprighting with microimplant anchorage is easier and independent of anchorage preparation on the anterior segment to avoid adverse effects such as premolar intrusion, which saves the clinician time.

Location of the TAD depends on the type of uprighting desired. If molar intrusion is indicated, the TAD can be inserted in the mandibular retromolar area—which usually offers adequate bone—then connected by an elastic thread passing occlusally to a mesial button on the molar (Fig 8-13a). To increase the intrusion effect, some composite can be bonded occlusally to take advantage of chewing forces. To upright a second molar impacted at the distal of the first molar, a bracket-head implant can be used with cantilever mechanics until a tube can be bonded (Fig 8-13b).

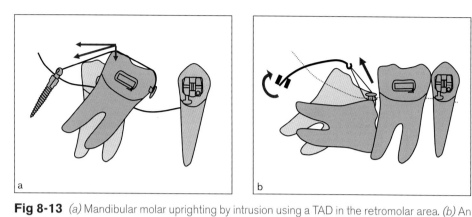

Fig 8-13 *(a)* Mandibular molar uprighting by intrusion using a TAD in the retromolar area. *(b)* An impacted molar can be uprighted with cantilever mechanics using a bracket-head TAD inserted in the retromolar area.

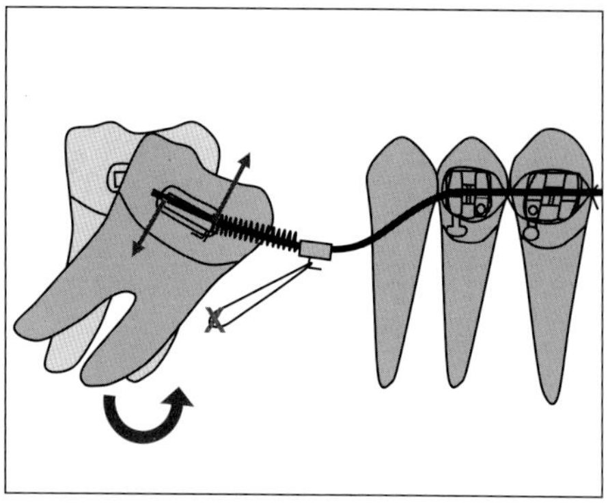

Fig 8-14 Mandibular molar uprighting to open space for prosthetic purposes.

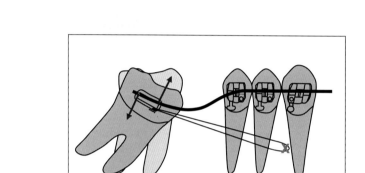

Fig 8-15 Mandibular molar uprighting using a microimplant to close space.

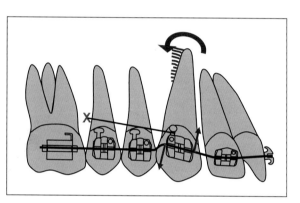

Fig 8-16 Microimplant inserted between the first molar and second premolar roots provides effective anchorage to prevent the rowboat effect.

If extrusion is not an issue, a TAD can be inserted buccally and mesially to the molar and force applied on a sliding hook and coil spring by means of a wire ligature. In fact, a straight wire passing through the molar tube creates a moment for uprighting; a distally directed force will increase the uprighting effect and help create enough room for future prosthetics (Fig 8-14).

If the space mesial to the molar is to be closed orthodontically, a microimplant should be inserted between the roots of the premolars or between the roots of the canine and first premolar. In this procedure, a laceback can be tied between the microimplant and the molar crown to prevent the latter from distalizing as the tooth uprights with straight wire (Fig 8-15). After the uprighting is accomplished, a chain elastic or closed coil spring can be applied between the microimplant and an auxiliary power hook to protract the molar.

In the maxilla, the tuberosity can be used as anchorage for molar uprighting with intrusion. This area, however, may not offer adequate bone for microimplant insertion unless longer and thicker microimplants, such as those offered by Dentos, are used.

Microimplants are excellent for providing anchorage support for some individual tooth movements and for preventing mechanical side effects from excessive inclination of the teeth, as in the rowboat effect. The rowboat effect is the tendency of a crown to move in an undesired direction owing to a couple. If a straight wire is inserted into an upright or distally tipped canine having angulated brackets, a counterclockwise moment occurs

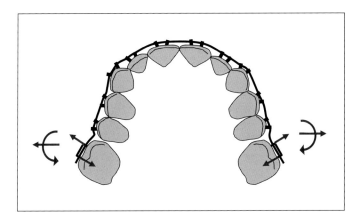

Fig 8-17 Transverse expansion is expected in mesiopalatally rotated molars when a straight wire is inserted in their tubes.

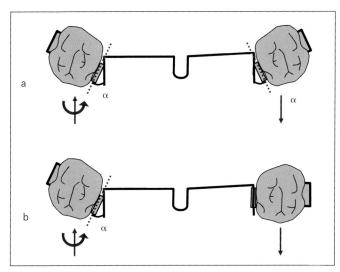

Fig 8-18 Molar rotation can be effectively corrected with a transpalatal arch. (*a*) Equal and opposite moments rule out balancing forces. (*b*) If the molars are initially rotated unequally, the balancing forces tend to move one molar mesially and the other distally. If this is not desired, the anchorage side should be reinforced by incorporating adjacent teeth. (α represents the amount of rotation of the molars relative to the sagittal plane.)

because of excessive bracket angulation, tending to move the crown mesially, with possible anchorage loss and flaring of the incisors.

To prevent the rowboat effect, the canine crown must be stabilized in place with strong anchorage. In cases where molar anchorage is critical, a microimplant is an excellent alternative to extraoral anchorage. The TAD is placed between the second premolar and first molar roots, and a laceback is applied directly to the canine bracket (Fig 8-16).

Molar Rotation

Rotation of maxillary molars is usually needed to gain space in the dental arch or to reinforce anchorage for canine and incisor retraction. The proper position of the maxillary first molar is key to a functional occlusion.[13] Most Class II cases need molar rotation to achieve a Class I relationship in the posterior segment. Also, correction of a rotated molar is needed to insert the facebow into the molar tube. Rotating the molars provides approximately 1 to 1.5 mm of space on each side of the dental arch. A straight wire engaged in mesiopalatally rotated molar tubes tends to expand the molars (Fig 8-17). Clinically, this expansion is not significant because the moment created by the flexible wire is not high enough to overcome the interdigitation. As archwire stiffness increases, transverse control of the molars becomes easier.

Molar rotation can also be achieved with a transpalatal arch, which can correct severe rotations better and more easily than a flexible main archwire. Figure 8-18 shows transpalatal arch mechanics in the correction of molar rotation. Note that equal and opposite moments create rotations with no balancing force on the system.

Conclusion

Molar distalizing procedures are best suited for correction of a Class II molar relationship in patients with transitional or early permanent dentition, a low mandibular plane angle, and a minimal or small mandibular arch length discrepancy. Cervical facebow headgear (often referred to as *Kloehn headgear*) used for maxillary molar distalization had been believed to create unfavorable effects such as extrusion and distal tipping of molars, distal rotation of occlusal and mandibular planes, and lengthening of the anterior lower face height.[14,15] However, research[16] based on patients treated by Silas Kloehn and John Kloehn showed that none of these effects occur when the patient is treated according to their protocol. Their system of treatment involves the use of a long outer bow that is adjusted every 6 to 8 weeks alternately up or down in relation to the occlusal plane. This procedure prevents the extrusion of maxillary molars and controls their axial inclination.

Patient compliance problems with headgear led to the development of several intraoral intra-arch appliances (eg, pendulum, Jones jig, and distal jet). These appliances were anchored on premolars and a palatal acrylic button (similar to the Nance appliance). While the use of these appliances did distalize and tip the molars, they also led to approximately equal amounts of anchorage loss.[17–19] In short, they were not as effective and useful as anticipated. TADs have been most useful in overcoming these difficulties.

Vertical height control in high-angle and open bite patients receiving orthodontic treatment is imperative because most orthodontic procedures have an extrusive effect on the molars. The vertical holding appliance (VHA),[20] which was developed at the University of Oklahoma, has been found to be helpful in controlling the extrusion of maxillary molars. The VHA is a simple palatal appliance anchored on the first molars supporting a 0.5-inch-diameter acrylic button in the middle of the palate between the molars. When the patient swallows, the tongue pressure on the button restricts the molar extrusion. Mandibular molar extrusion can also be controlled by application of a fixed lingual arch.[21]

References

1. Andrews LF. The six keys to normal occlusion. Am J Orthod 1972;62:296–309.
2. Kim TW, Artun J, Behbehani F, Artese F. Prevalence of third molar impaction in orthodontic patients treated nonextraction and with extraction of 4 premolars. Am J Orthod Dentofacial Orthop 2003;123:138–145.
3. Kandasamy S, Woods MG. Is orthodontic treatment without premolar extractions always non-extraction treatment? Aust Dent J 2005;50:146–151.
4. Kloehn SJ. A new approach to the analysis and treatment in mixed dentition. Am J Orthod 1953;39:161–186.
5. Lee JS, Kim JK, Park YC, Vanarsdall RL Jr. Applications of Orthodontic Mini-Implants. Chicago: Quintessence, 2007:195–197,203.
6. Graber TM. Current Orthodontic Concepts and Techniques. Philadelphia: Saunders, 1969:919–988.
7. Tosun Y. Comparative Study of the Effects of Cervical and High Pull Headgears on Dento-Cranio–Facial Structures of Class II/1 Cases in the Transitional Dentition [thesis]. İzmir, Turkey: Aegean Univ, 1989.
8. Gianelly AA, Bendar J, Dietz VS, Koglin J. An approach to nonextraction treatment of Class II malocclusions. In: Nanda R (ed). Biomechanics in Clinical Orthodontics. Philadelphia: Saunders, 1997:258.
9. Locatelli R, Bendar J, Dietz VS, Gianelly AA. Molar distalization with superelastic NiTi wire. J Clin Orthod 1992;26:277–279.
10. Stepovich ML. A clinical study on closing edentulous spaces in the mandible. Angle Orthod 1979;49:227–233.
11. Tosun Y. Biomechanical Principles of Fixed Orthodontic Appliances. İzmir, Turkey: Aegean University, 1999:224.
12. Aras A, Tosun Y. Application of "Memory Titanol Spring" in the first molar extraction space closure. Turk J Orthod 1999; 12:141–148.
13. Burstone CJ. How to level the occlusal plane in deep bite cases. Presented at the 4th International Orthodontic Congress, San Francisco, 12–17 May 1995.
14. Wieslander L. The effect of orthodontic treatment on the concurrent development of the craniofacial complex. Am J Orthod 1963;49:15–27.
15. Merrifield LL, Cross JJ. Directional forces. Am J Orthod 1970; 57:435–464.
16. Hubbard G, Nanda RS, Currier GF. A cephalometric evaluation of nonextraction cervical headgear treatment in Class II malocclusions. Angle Orthod 1994;64:359–370.
17. Ghosh J, Nanda RS. Evaluation of an intraoral maxillary molar distalization technique. Am J Orthod Dentofacial Orthop 1996; 110:639–646.
18. Brickman, CD, Sinha PK, Nanda RS. Evaluation of the Jones jig appliance for distal molar movement. Am J Orthod Dentofacial Orthop 2000;118:526–534.
19. Ngantung V, Nanda RS, Bowman SJ. Posttreatment evaluation of the distal jet appliance. Am J Orthod Dentofacial Orthop 2002;120:178–185.
20. DeBerardinis M, Stretesky T, Sinha PK, Nanda RS. Evaluation of the vertical holding appliance in treatment of high-angle patients. Am J Orthod Dentofacial Orthop 2000;117:700–705.
21. Villalobos FJ, Sinha PK, Nanda RS. Longitudinal assessment of vertical and sagittal control in the mandibular arch by the mandibular fixed lingual arch. Am J Orthod Dentofacial Orthop 2000;118:366–370.

CHAPTER 9

Space Closure

Extraction is a common approach in orthodontic treatment to correct arch-to-tooth size discrepancies and skeletal problems. A number of factors should be considered before determining the need for extraction. The most important factors are:

- Severity of crowding
- Vertical growth pattern
- Midline discrepancies
- Incisor-lip relationship
- Anchorage

Space closure is one of the most important steps in treatment after extraction. The strategy of space closure should be based on a careful diagnosis and treatment plan made according to the specific needs of the individual. No matter the technique used, the extraction space can be closed in three ways:

- Retraction of anterior teeth (maximum anchorage)
- Protraction of posterior teeth (minimum anchorage)
- A combination of both (moderate anchorage)[1–4]

Space closure with conventional intra-arch mechanics is a sort of tug-of-war between anterior and posterior segments. According to Newton's third law, to every action there is always an equal and opposite reaction; therefore, for all distally directed forces (anterior retraction), the reaction will be mesially directed (posterior protraction). If maximum anchorage is indicated, the entire[1]–or at least 75% of the extraction space–needs to be closed by retraction of anterior teeth.[2–4] In this case, there are several possible mechanisms to close these spaces:

- Retraction of anterior teeth without involving the posterior teeth by means of extraoral support such as J-hook headgear or microimplant (temporary anchorage devices [TADs]) mechanics
- Retraction of anterior teeth without involving the posterior teeth by taking support from the other arch (ie, Class II elastics)
- Neutralizing the mesial forces acting on the posterior teeth with full-time headgear wear
- Applying retraction force on the anterior teeth while headgear is worn
- Using differential mechanics

As far as maximum anchorage is concerned, excellent patient cooperation is usually necessary to obtain the proper interarch relationship. In the first four options (unless microimplant anchorage is used in the first option), the success of treatment is mainly dependent on patient compliance. Maximum anchorage is quite difficult to achieve unless intraoral differential tooth movement strategies are applied. Naturally, headgear or other anchorage auxiliaries such as transpalatal arch, lingual stabilizing arch, and Class II elastic support might be needed to obtain effective retraction.

In a maximum anchorage case, the timing of extraction is also important because posterior teeth can easily drift mesially after extraction. In some Class II, division 1 patients, for instance, it is necessary to wait until leveling is completed to extract the appropriate tooth and start the retraction process.

Differential Mechanics

Maximum anchorage

To achieve maximum anchorage, the mesial forces on the posterior teeth must be reduced or neutralized with differential moments, which use the difference in anchorage between the anterior and posterior teeth.

Classically, a tooth tips when a simple force is applied but resists if a couple is added to the system. As the moment increases, the moment-to-force (M/F) ratio, and thus the anchorage of the tooth, increases; therefore, the force system applied to the tooth defines its anchorage value based on more than simply the number and total surface of its roots. Occasionally, depending on the force system applied, the anchorage of a single-rooted tooth can be higher than that of a multirooted tooth.[5] To obtain maximum anchorage in the posterior segment, the M/F ratio on the molars should be high enough to effect root movement. This can be achieved either by increasing the clockwise moment or reducing the mesially directed force acting on the molar. An M/F ratio of approximately 12:1 and higher results in root movement and a distal crown-tipping tendency of the posterior teeth, resisting the mesially directed force. Because the amount of force applied to each segment is the same, one should adjust the respective amount of moment on each side to get differential tooth movement. The M/F ratio on the anterior segment should be as low as 6:1 to obtain controlled tipping. Clinically, because controlled tipping is quicker and easier than root movement, anterior retraction can be achieved before posterior protraction occurs (Fig 9-1). Even though this mechanism works effectively, on a practical basis, nighttime headgear wear is usually needed to control molar inclination and neutralize the vertical force vector.

Differential moments are balanced by vertical forces, which are intrusive on the anterior and extrusive on the posterior segment. The amount of balancing force depends mainly on the strength of the moments and the interattachment distance. Large moments result in large vertical forces, which may be detrimental to the vertical dimension of the face, particularly in high-angle patients. As the distance between segments increases, the amount of vertical force decreases and vice versa.

Moderate anchorage

In moderate anchorage, the extraction space needs to be closed by an equal amount of retraction of anterior teeth and protraction of posterior teeth.[1–4] When differential mechanics are used, the moments applied to the anterior and posterior segments must be equal and opposite. To obtain controlled tooth movement, an M/F ratio of approximately 6:1 to 8:1 should be applied to each segment. After the extraction space closes, higher moments are needed to obtain root movement.

Minimum anchorage

If minimum anchorage is indicated, all[1]—or at least 75%—of the extraction space needs to be closed by protraction of the posterior teeth.[2–4] This is nearly the opposite of maximum anchorage mechanics.

When differential mechanics are used to achieve minimum anchorage, the distally directed force on the anterior teeth must be reduced or neutralized with differential moments. To obtain maximum anchorage on the anterior segment, the M/F ratio on the incisors should be high enough to obtain root movement. This can be achieved by either increasing the clockwise moment or reducing the distally directed force acting on the anterior teeth. An M/F ratio of approximately 10:1 to 12:1 results in root movement and forward crown tipping of the incisors, resulting in resistance against the distal force. The M/F ratio at the posterior segment should be as low as 6:1 to obtain controlled tipping. Clinically, space closure with differential anchorage mechanics is not convenient. The anchorage of the anterior segment

Fig 9-1 Example of force system for space closure with maximum anchorage. High M/F ratio on the posterior segment and lower M/F ratio on the anterior segment results in root movement of the molars and premolars and controlled tipping of the incisors and canines.

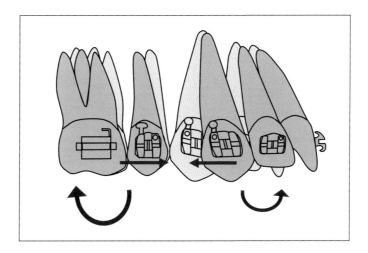

Fig 9-2 *(a and b)* In minimum anchorage frictional mechanics, the posterior teeth should be moved one by one with intermaxillary elastic support to avoid undesired retraction of anterior teeth.

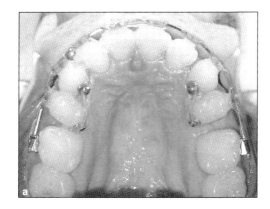
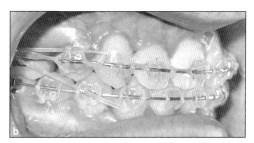

usually needs to be reinforced (eg, with a facemask worn at night) or the posterior teeth may need to be protracted one at a time because incisor anchorage is not strong enough for en masse protraction (Fig 9-2). As explained in the section "Molar Protraction" in chapter 8, the convenient way to close extraction spaces in a minimum anchorage case is to move the premolars and molars one at a time with push coils while the anchorage is supported with intermaxillary elastics.

General Strategies in Space Closure

Two basic biomechanical strategies can be used to close extraction spaces: frictional and frictionless. In either system, the tooth movements can be two-staged (canine distalization and incisor retraction) or en masse. Canine distalization is necessary in anterior crowding cases wherein round tripping (ie, protrusion and retrusion of incisors) is to be avoided. Thanks to flexible retraction archwires, use of microimplant anchorage mechanics, and new concepts in biomechanics, en masse retraction can be applied in most extraction cases.

Space closure in frictional mechanics

In frictional mechanics, anchorage is a serious concern because a significant part of the force applied to tooth movement is lost in the friction between bracket, ligature, and wire. Therefore, particularly in patients in whom anchorage is critical, use of headgear is more important than in the frictionless system. Space closure in frictional mechanics has usually been performed in two stages to avoid straining the anchorage teeth; however, this technique is usually more time-consuming than one-stage (en masse) retraction,[6] and it places more strain on anchorage than commonly recommended.

9 | Space Closure

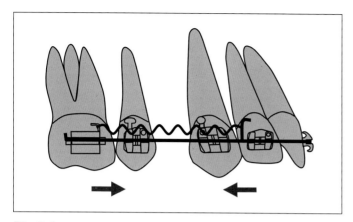

Fig 9-3 In en masse retraction with frictional mechanics, the archwire slides through the brackets and tubes of the posterior teeth.

Two-stage space closure
Canine distalization

In frictional mechanics, the canines move distally by sliding on the archwire. As long as the applied force is continuous and optimum (approximately 200 g),[6] tooth movement is achieved by slight tipping and uprighting cycles (see Fig 4-4). Several factors related to friction such as amount of force, angle between wire and bracket, wire size, wire stiffness, wire and bracket materials, and ligature material and tightness can affect the movement.[5] Generally, the force needed to overcome friction is of the same magnitude as the force needed to move teeth.[7,8] With a given optimum force, a wire with high elasticity (less stiff) would deflect more than a stiff wire and would cause more tipping. On the other hand, a similar effect could be obtained with a stiffer wire–excessive force combination. Excessive force causes a deep bite by incisor extrusion, anchorage loss due to binding, and permanent wire deformation (see Fig 4-9). Stiff, round wire–wire ligature–light force seems to be the optimum combination for efficient canine retraction. Adjusting the tightness of the ligature (see chapter 4) is also recommended to avoid binding and related side effects.

Incisor retraction

After canine distalization, the incisors are retracted to close residual spaces. Incisor retraction in sliding mechanics requires stronger anchorage compared with canine distalization. Incorporating the posterior teeth (including canines) is generally not enough for strong anchorage. For moderate anchorage, chain elastics work relatively well, but for maximum anchorage, headgear and/or Class II elastic support is usually needed.

En masse retraction

As in other sliding mechanics, en masse retraction in frictional mechanics requires strong anchorage support. In this technique, as the anterior teeth move distally, the archwire slides through the brackets and tubes of the posterior teeth (Fig 9-3). The interaction between the wire and tubes may result in high friction—even binding—causing loss of anchorage. To avoid friction, particularly that due to third-order interaction, the edges of rectangular archwires should be rounded with a diamond bur and polished with a rubber wheel (see Fig 8-7). For effective sliding, there must be a 0.002-inch play between wire and bracket slot. It is therefore suitable to use 0.016 × 0.022–inch wire in a 0.018-inch slot. Practically, to avoid friction, the thickest leveling wire can be kept in the mouth for 1 or 2 months until enough molar and premolar torque has been attained. If a thinner rectangular wire with rounded edges is engaged in the brackets, third-order interaction would be eliminated, but this could result in the posterior teeth moving mesially. In patients in whom anchorage is critical, retraction can be safely achieved with intramaxillary elastics along with headgear or intermaxillary elastics only. Owing to the friction factor, the differential tooth movement principle cannot be applied to en masse retraction in frictional mechanics, but it does work in minimum to moderate anchorage cases. Nickel titanium (NiTi) coil springs between molars and canines provide optimum force and tooth movement. Class II elastics support the anchorage in reaching a Class I molar relationship.

En masse retraction with microimplant mechanics

En masse retraction, which is one of the most challenging procedures when using conventional techniques, can be performed more efficiently with microimplant anchorage mechanics. With traditional techniques, en masse retraction can be done only against extraoral anchorage, which is dependent on patient compliance.

In en masse retraction using microimplant anchorage, the location of the TAD is determined according to the

Fig 9-4 Force applied between the anterior power hook and a TAD inserted between the second premolar and first molar roots 8 to 10 mm from the archwire (medium region, *gray band*) *(a)* causes en masse retraction of the anterior teeth with translation *(b)*. (Reprinted from Sung et al[9] with permission.)

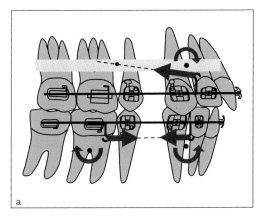 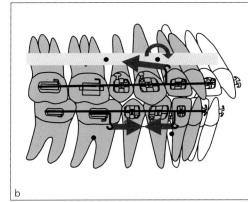

Fig 9-5 Force applied between the anterior power hook and a TAD inserted between the second premolar and first molar roots less than 8 mm from the archwire (below *gray band*) *(a)* causes en masse retraction of the anterior teeth combined with clockwise rotation, which helps correct anterior open bite *(b)*. (Reprinted from Sung et al[9] with permission.)

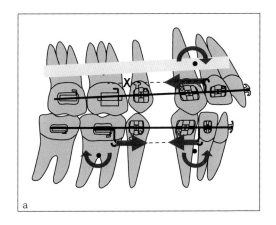 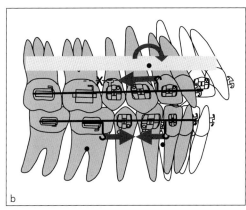

Fig 9-6 Force applied between the anterior power hook and TAD inserted between the second premolar and first molar roots more than 10 mm from the archwire (above *gray band*) *(a)* causes en masse retraction of the anterior teeth combined with clockwise rotation, which helps correct anterior deep bite *(b)*. (Reprinted from Sung et al[9] with permission.)

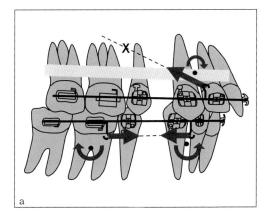 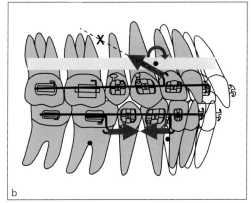

patient's need. Sung et al[9] have defined three regions according to the direction of force: low, medium, and high. A TAD inserted between the maxillary first molar and second premolar roots 8 to 10 mm from the archwire would be considered in the medium region. Above 10 mm is high and below 8 mm is low. A coil spring force applied to a 6-mm-long power hook between the maxillary lateral incisor and canine passing through the medium region would likely cause translation of the anterior teeth (Fig 9-4). A force passing through a TAD inserted in the low region would cause the maxillary dentition to rotate clockwise and extrude, which is helpful in correcting anterior open bite (Fig 9-5). In contrast, if the microimplant is in the high region, the anterior teeth would intrude, which is helpful in correcting a deep bite (Fig 9-6).

Space closure in frictionless mechanics

Space closure in frictionless mechanics can be achieved by either a looped, continuous archwire or the segmented arch technique.

Continuous archwire

The continuous archwire technique is performed with a closing loop bent on the main archwire. No matter the technique, the spring characteristics of a closing loop–a critical part of its design–are largely determined by three factors: size of the archwire, configuration of the loop, and distance between points of attachment (interbracket distance). The principal means of increasing the effectiveness of loops are:

- Incorporating more wire in the loop (eg, adding a helix)
- Using a smaller size wire
- Increasing the interbracket distance
- Changing the configuration of the loop

The M/F ratio of a loop increases as the amount of wire incorporated gingivally increases.[10,11] The longer the interbracket distance, the higher the working range of a loop. Equivalent properties can be obtained by increasing the amount of wire as the size increases and by using simpler loops as the wire size decreases. The proper selection of the combination depends on the comparative clinical risks and benefits. Placing more wire in a loop increases its working range and increases the M/F ratio, but more wire might be uncomfortable for the patient. On the other hand, simple loops (ie, vertical loops) are comfortable, but the M/F ratio is considerably lower.

The average amount of activation recommended for a vertical loop such as a Bull, Sandusky, or delta is about 1 mm per month. A Bull loop in an 0.018 × 0.025–inch stainless steel (SS) wire activated 1 mm can deliver approximately 500 g.[1] The legs of the loop close quickly following activation, which is clinically fail-safe.[6] Proffit[6] suggests that tooth movement be stopped after a prescribed range of movement to avoid mechanical side effects, in case the patient fails to present for his or her scheduled visit.

The M/F ratio produced by a loop can be increased by appropriate gable bends placed in the legs (Fig 9-7). These bends generate root-paralleling moments after the extraction spaces have been closed. The angle of the bend depends mainly on the size or stiffness of the wire, that is, 30 to 40 degrees for 0.016 × 0.022–inch SS, 20 to 30 degrees for 0.017 × 0.025–inch SS, and above 40 degrees for 0.017 × 0.025–inch titanium-molybdenum alloy (TMA). The angle of the bend also depends on the type of retraction and anchorage needs of the case. If the incisors are to be retracted with root movement, the gable bend should be increased. The anterior moment can be increased, if necessary, by placing palatal root torque in the wire.

A gable bend approximates a V-bend; therefore, its position between segments affects the mechanics of tooth movement (see chapter 3). Moving the loop 1 or 2 mm toward one side usually creates a moment differential.[2] In a moderate anchorage case, the loop should be positioned in the middle of the segments to obtain equal and opposite moments. In a maximum anchorage case, placing the loop distally produces a higher moment; however, this is limited by the interbracket distance at the extraction site. The loop moves backward as the wire is cinched behind the molar tube; therefore, in a first premolar extraction case, the loop should be placed initially more anteriorly to ensure that it will not interfere with the second premolar bracket. As the loop moves backward, the posterior moment increases.

When the loop is placed too far anteriorly, combined with palatal root torque on the incisors, the anterior moment becomes higher than the posterior. This difference of moment is balanced with extrusive force on the incisors and intrusive force on the molars. Class II elastics to reinforce the anchorage may cause extrusion of the incisors and be detrimental to the treatment result. If anchorage is critical, headgear with or without a transpalatal arch is the method of choice to reinforce it.

Segmented arch mechanics

In the segmented arch technique, the anterior and posterior segments are converted into two big "teeth" by means of rectangular wires and a transpalatal arch. Therefore, any mechanics used in this technique are considered a tug-of-war between two teeth, as has been explained earlier in this chapter. Space closure is performed with a 0.017 × 0.025–inch TMA T-loop wire inserted into the molar tubes and a crimpable vertical

General Strategies in Space Closure

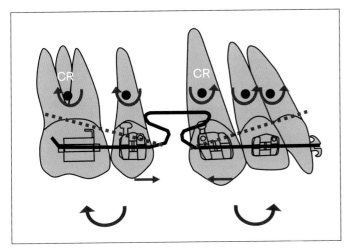

Fig 9-7 In space closure with continuous archwire mechanics, a gable bend should be placed in the mesial and distal legs of the loop to obtain root movement. Because the archwire is passed through the brackets, the teeth move around their own centers of rotation (CR).

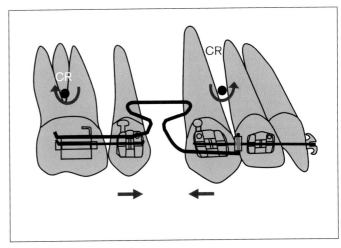

Fig 9-8 In space closure with the segmented arch technique, the anterior and posterior segments can be considered two big "teeth," which move around their respective centers of rotation (CR). (Redrawn with permission from Melsen et al.[10])

tube inserted between the lateral incisor and canine brackets.

The segments move around their respective centers of rotation[10] (Fig 9-8). The amount of activation of the loop and its position between the two attachments defines the biomechanical design. Nanda and Kuhlberg[2] and Burstone[11] suggest activating the T-loop 4 to 6 mm, depending on the specific needs of the patient, to obtain controlled tooth movement. When the T-loop (or V-bend) is at the midpoint of the two attachments, it creates equal and opposite moments on each side. When it moves off center, it generates a higher moment on the nearer side.[2,4,12–16] Unlike continuous archwire mechanics, the long distance between attachments makes it possible to position the loop (V-bend) posteriorly or anteriorly, depending on the anchorage needs of the patient.

In maximum anchorage cases, the extraction spaces are closed mainly with the retraction of the anterior teeth, with little or no forward movement of the posterior teeth. Application of differential mechanics helps support posterior anchorage, but usually not enough to accomplish en masse retraction. Therefore, headgear and/or Class II elastics are usually needed to reinforce the anchorage (Fig 9-9).

Segmented arch mechanics offer effective anterior retraction in maximum anchorage cases because both the TMA wire and the design of the T-loop help reach high M/F ratios when needed. In maximum anchorage cases, differential anchorage is obtained when the posterior moment is higher than the anterior moment. In differential anchorage, the goal is to obtain root movement of the molars and tipping of the incisors. Root movement of the posterior segment can be obtained in two ways: increasing the bend on the posterior leg of the T-loop (increasing the posterior moment)[17] and positioning the T-loop close to the posterior attachment.

An M/F ratio of approximately 10:1 to 12:1 generates root movement of the posterior teeth. The effect of the moment is to tip the molar crowns distally, thus resisting against mesially directed forces, while the anterior segment is retracted with controlled tipping by an M/F ratio of approximately 6:1 to 7:1.

Differential moments cause vertical balancing forces, which are intrusive at the anterior and extrusive at the posterior segments. Molar extrusion must be controlled in high-angle patients to avoid mandibular rotation. High-pull headgear may be needed to control extrusion and inclination of the molars and correct the cant of the posterior segment.

In moderate anchorage cases, the extraction spaces are closed with equal amounts of movement of the anterior and posterior segments. If the loop is positioned

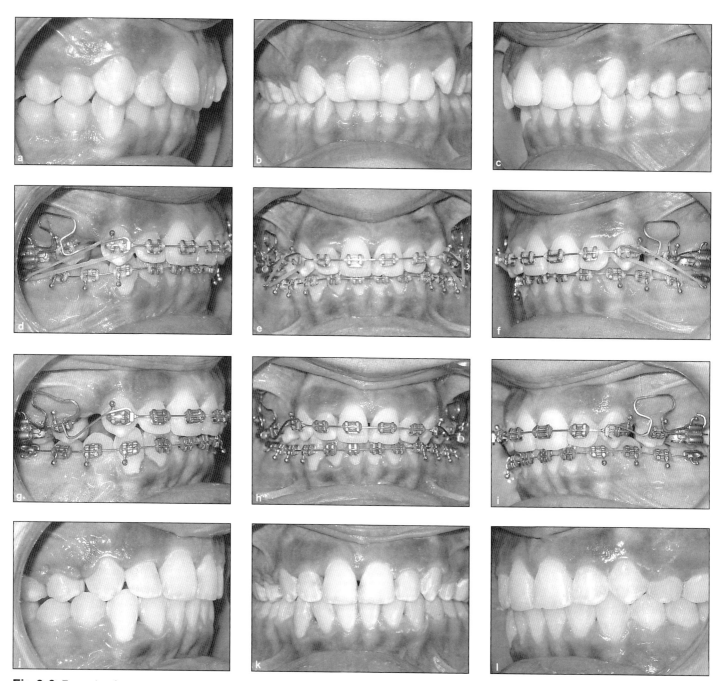

Fig 9-9 Example of en masse space closure with segmented arch mechanics using 0.016 × 0.022–inch TMA T-loops and Class II elastics. *(a to c)* Before treatment. *(d to f)* Start of retraction with T-loop combined with Class II elastics. *(g to i)* After 3 months of treatment. Note the improvement of the canines' inclinations despite retraction force, due to high M/F ratio delivered by T-loops. *(j to l)* Completion of treatment following 5 months of retraction mechanics.

centrally between the segments, equal and opposite moments occur; therefore, there are no balancing forces on the system. Nanda and Kuhlberg[2] suggest activation of a TMA T-loop up to 6 mm to obtain an M/F ratio of approximately 6:1 on each side. As the loop deactivates, the M/F ratio increases up to 27:1, producing three phases of tooth movement—tipping, translation, and root movement.[2,11] Proffit[6] suggests that segmented arch mechanics are not fail-safe; arch form and the proper vertical relationship cannot be controlled if the retraction loop is distorted or activated incorrectly. Nanda and Kuhlberg[2] suggest that space closure be monitored periodically and that the spring be reactivated when the occlusal planes regain parallelism.

In minimum anchorage cases, space closure is accomplished mainly with mesial movement of the posterior segment. Protraction of the molars and premolars with differential mechanics is fairly difficult to accomplish because anterior anchorage is not strong enough and side effects are usually unavoidable.

To obtain a differential, the anterior moment should be increased and the loop placed close to the anterior segment. Differential moments generate vertical balancing forces, which are extrusive on anteriors and intrusive on posteriors. The major side effect is a deep bite due to incisor extrusion and retrusion. In the maxillary dentition, the facemask in combination with Class III elastics increases the mesial force on the molars. In the mandibular dentition, Class II elastics may be needed to reinforce anterior anchorage. To minimize side effects, the T-loop should be activated 1 to 2 mm. In minimum anchorage cases, second premolar extraction is helpful to obtain higher anchorage in the anterior segment. If both maxillary and mandibular dental arches need minimum anchorage, a facemask to the maxillary dentition combined with Class II elastics is needed to support anterior anchorage and avoid retrusion of the maxillary and mandibular incisors.

References

1. Ulgen M. Treatment Principles in Orthodontics, ed 3. Ankara, Turkey: Ankara Univ, 1990.
2. Nanda R, Kuhlberg A. Biomechanical basis of extraction space closure. In: Nanda R, Burstone CJ. Retention and Stability in Orthodontics. Philadelphia: Saunders, 1993:156–187.
3. Burstone CJ, Van Steenbergen E, Hanley KJ. Modern Edgewise Mechanics: Segmented Arch Technique. Farmington, CT: Univ of Connecticut, 1995:5–16.
4. Tosun Y. Biomechanical Principles of Fixed Orthodontic Appliances. İzmir, Turkey: Aegean University, 1999:195.
5. Mulligan TF. Common Sense Mechanics in Everyday Orthodontics. Phoenix: CSM, 1998.
6. Proffit WR. Contemporary Orthodontics. St Louis: Mosby, 1986:428,433.
7. Andreasen GF, Quevedo FR. Evaluation of frictional forces in the 0.022" × 0.028" edgewise bracket in vitro. J Biomech 1970; 3:151–160.
8. Peterson L, Spencer R, Andreasen GF. A comparison of frictional resistance for Nitinol and stainless steel wires in edgewise brackets. Quintessence Int Dent Dig 1982;13:563–571.
9. Sung JH, Kyung HM, Bae SM, Park HS, Kwon OW, McNamara JA Jr. Microimplants in Orthodontics. Daegu, Korea: Dentos, 2006:15–32.
10. Melsen B, Fotis V, Burstone CJ. Vertical force considerations in differential space closure. J Clin Orthod 1990;24:678–683.
11. Burstone CJ. The segmented arch approach to space closure. Am J Orthod 1982;82:361–378.
12. Burstone CJ, Koenig HA. Creative wire bending—The force systems from step and V bends. Am J Orthod Dentofacial Orthop 1988;93:59–67.
13. Ronay F, Kleinert W, Melsen B, Burstone CJ. Force system developed by V bends in an elastic orthodontic wire. Am J Orthod Dentofacial Orthop 1989;96:295–301.
14. Demange C. Equilibrium situations in bend force systems. Am J Orthod Dentofacial Orthop 1990;98:333–339.
15. Bequain DMJ. Etude mécanique de la coudure d'un fil. Orthod Fr 1994;65:547–557.
16. Deblock L, Petitpas L, Ray B. Mécanique de recul incisif maxillaire. Orthod Fr 1995;66:667–685.
17. Staggers JA, Germane N. Clinical considerations in the use of retraction mechanics. J Clin Orthod 1991;25:364–369.

Glossary

binding During tooth movement along an archwire, a high couple occurs between bracket and wire, owing to overtipping of the tooth. Binding is the locking of these two materials due to friction. As a result of binding, tooth movement stops. Binding can also occur during sliding of the wire in en masse movements.

bowing effect Increase of overbite by incisor extrusion after a straight wire is placed in incisor brackets and angulated canine brackets.

breaking point The point at which a wire material fails.

center of rotation The imaginary point around which a tooth rotates. The location of this point changes depending on the force system applied to the tooth (ie, the moment/force ratio).

corrosion Loss of weight of a metal and change of its mechanical properties under various chemical influences.

couple A system having two equal and parallel forces acting in opposite directions. Every point on the object to which a couple is applied is under a rotational influence in the same direction and magnitude. The object rotates about its center of resistance regardless of where the couple is applied. For example, torque applied to a bracket makes the tooth rotate around its center of resistance and not around the bracket.

differential force (anchorage) principle Taking advantage of the anchorage differences between two teeth (or groups of teeth) when moving them toward or away from each other.

dynamic (kinetic) friction Friction that exists during movement of an object; the amount of force the object must overcome to continue moving.

elastic deformation The shape deformation of a wire that occurs up to its elastic limit.

elastic material Material that can return all the energy it absorbs, thus regaining its original dimensions when the stress is removed.

fatigue Weakening of a material under repeated stress.

force The factor that causes an object to change its shape or position in space. Force is a vector having a line of action, direction, magnitude, and point of application.

formability The amount of permanent deformation a wire can sustain before failure.

free object analysis Analysis of an isolated part of a force system representing the whole system or an object in static equilibrium. This analysis helps one see the entire perspective of the object or system.

frictional force The resistance that occurs at the contact surface in the opposite direction to the movement when two objects in contact are forced to move along each other.

Hooke's law According to this law, stress applied to a material up to its elastic limit is directly proportional to the strain.

hysteresis The difference between the slope formed when a superelastic wire is activated and transforms from the austenitic phase to the martensitic phase and that formed when the force is removed and the wire follows a different path below that of activation, returning to the austenitic structure.

laceback A ligature tied between the molar tube and canine bracket to prevent the canine crown from moving mesially.

limit of elasticity The point beyond which permanent deformation occurs.

modulus of resilience The area in a stress-strain diagram between the starting point, elasticity limit, and vertical line from the elasticity limit to the *x*-axis.

modulus of toughness The area in a stress-strain diagram under the slope brought down to the *x*-axis from the starting point and the failure point.

moment The magnitude of force (F) times perpendicular distance (d) drawn from the center of resistance to the line of action of the force, shown as $M = F \times d$.

moment/force ratio The ratio of the moment of a couple effective on the tooth and the force applied to the crown.

normal force The force perpendicular to the surface that causes friction between two objects.

permanent deformation The permanent shape deformation of a wire due to excessive stress beyond the elastic limit.

pitch angle The angle between a line perpendicular to the long axis of a spring and the inclination of the windings of the spring.

plastic material A material that does not return to its original dimensions when stress is removed.

pure rotation The rotation of an object that occurs because of a couple.

relaxation The decrease or total reduction of force over time, occurring under constant stress.

rotation movement Movement of any straight line on a body by changing its angle relative to a fixed reference frame.

round tripping Bringing a tooth back to its original position after moving it. One of the causes of root resorption.

rowboat effect The tendency of a tooth crown to move in an undesired direction because of a couple. In the straight-wire technique, if the canines are too upright due to the second-order angulation of the canine bracket, a moment occurs when an archwire is inserted, tending to move the crown mesially. The rowboat effect may result in anchorage loss and flaring of the incisors.

shape memory The shape "remembering" process by which an alloy returns to its original shape after it is heated above a certain transition temperature.

springback The distance on the *x*-axis of a stress-strain diagram between 0.1% permanent deformation and the elastic limit of a wire that has been deflected and let go.

static The branch of physics concerning the behavior of objects in balance while under the effect of a force.

static frictional force The amount of force necessary for an object to start moving.

stiffness The amount of resistance a wire exhibits against bending and drawing (ie, the amount of force needed to bend or draw a wire).

strength The total force application capacity of a wire material.

superelasticity The characteristic of a wire that delivers the same amount of force independent of the amount of activation.

surface hardness The resistance of a material against localized and continuous pressure. The Brinell and Rockwell hardness indexes are the most frequently used surface hardness scales in engineering.

transition temperature The temperature at which transition of NiTi alloy occurs between its austenitic and martensitic crystalline structures and vice versa.

translation (bodily) movement The type of movement of a body on any straight line without changing its angle relative to a fixed reference frame.

ultimate strength limit The ultimate strength a wire material can resist against an increasing stress.

vector A physical quantity having a point of application, magnitude, line of action, and direction.

working range The maximum amount a wire can be deformed without exceeding its material limit. This is important in orthodontic practice as it shows the working distance that a wire can be deformed within its elastic limit.

Young's modulus The slope of a stress-strain curve of a wire below its elastic limit.

Index

Page numbers followed by "f" denote figures; those followed by "t" denote tables

A

Acceleration, law of, 1
Action and reaction, law of, 1
Alumina, 75
Alveolar support, loss of, 12–13, 13f
Anchorage control
  definition of, 83
  extraoral appliances for
    adjustment to direction of force, 87–89, 88f
    anteroposterior force application, 86
    asymmetric force application, 91–95
      asymmetric arms, 91–93, 92f
      canine hooks, 93, 92f
      reverse headgear, 93–95, 93f, 94f
    categorization of, 86
    entire dental arch application of, 90–91, 91f
    frontal plane analysis, 89, 89f
    objectives of, 86
    sagittal plane analysis, 86–87, 87f
      cervical headgear, 86, 87f
      high-pull headgear, 86–87, 87f
    transverse plane analysis, 90, 90f
  intraoral methods of
    cortical bone anchorage, 84, 85f, 95
    increase the number of teeth, 83, 84f
    lip bumper, 85, 85f
    Nance appliance, 83–84, 136
    sliding jig, 84
    transpalatal arch, 84–85, 89
    uprighting springs, 84, 84f
  maximum, 145–146, 147f, 151
  microimplants. See Microimplants.
  minimum, 145–146, 153
  moderate, 145–146, 151, 153
  temporary devices for
    deep bite correction using, 106, 107f
    description of, 95–96
    microimplants. See Microimplants.
    molar distalization using, 135, 135f
    molar intrusion with, for skeletal open bite correction, 122–123, 123f
    transpalatal arch and, 123f
Angulated curve of Spee, 110, 110f, 112f
Anterior crossbite, 126f
Anterior teeth, retraction of, 145. See also Incisor(s).
Anteroposterior discrepancies, 133–144
Anteroposterior force, 86
Antitip bend, 65
Arch. See specific arch.
Archwire, continuous, 150, 151f. See also Wire(s).
Artistic bends, 63, 64f
Asymmetric force, 91–95
Austenite, 28

B

Begg appliance, 14
β-titanium wires
  characteristics of, 31
  indications for, 37
  stiffness of, 20–21, 22t, 36t
  torsional properties of, 31, 32t
  working range of, 22t
Binding, 74, 77
Bite, 69, 69f. See also Crossbite; Deep bite; Open bite.
Bowing effect, 69, 69f
Bracket(s)
  edgewise, 76
  friction and, 75
  milled, 76
  self-ligating, 77f
  types of, 49–51, 50t
  width of, 75–76
Bracket angulation
  description of, 57, 59f
  straight-wire, 65, 67–68
Bracket-head screw, 130, 130f
Bracket-wire angle, 77–78
Bracket-wire friction
  analysis of, 73–78
  factors that affect, 75b, 75–78
  saliva effects on, 76
  types of, 71
Brittle, 19
Buccinator muscle, 85
Bull loop, 150
Burstone geometry classes, 57–59, 58f

C

Canine distalization
  force amount for, 27, 74
  friction during, 73, 74f
  in Class II, division 2 deep bite correction, 112
  in space closure, 148
  incisors after
    extrusion of, 79, 79f
    retraction of, 148
  stages of, 74f
Canine hooks, 92f, 93
Cantilever beam test, 24, 24f
Cantilevers, 38f–40f, 38–39
Canting of maxillary occlusal plane, 118, 119f
Center of mass, 6
Center of resistance
  description of, 5–6, 6f–7f
  loss of alveolar support effects on, 12, 13f
Center of rotation
  description of, 6
  moment-to-force ratio effects on, 9f, 11
Ceramic brackets, 50–51
Cervical headgear, 86, 87f, 108, 134, 143
Chain elastics. See Elastomeric chains.
Chinese NiTi, 29
Chloride, 26
Chrome-cobalt coil springs, 40–41, 41f
Class I geometry, 57, 58f
Class I molar relationship, 133
Class II elastics, 45–47, 46f–47f, 46t, 92f, 137, 152f, 153
Class II geometry, 57, 58f
Class II molar relationship, 133
Class III elastics, 45, 47
Class III geometry, 58f, 59
Class IV geometry, 58f, 59
Class V geometry, 58f, 59
Class VI geometry, 58f, 59
Closure of space. See Space closure.
Coefficient of static and kinetic friction, 73
Coil springs
  chrome-cobalt, 40–41, 41f
  description of, 18, 39–41, 41f
  elastics vs, 48–49
  molar distalization with, 135f, 135–136
  Nance appliance and, 136, 137f
  nickel titanium, 40–41, 41f
  stainless steel, 40–41, 41f
Combined headgear, 87
Compression, 17
Consistent mechanics, 38f, 39
Continuous archwire technique, for space closure, 150, 151f
Continuous forces, 4, 5f, 32
Continuous intrusion arch, 105, 105f–106f
Controlled tipping, 8–9, 9f, 66, 146
Corrosion, 26–27

Cortical bone anchorage, 84, 85f, 95
Couple, 6–7, 7f, 18
Crevicular corrosion, 26
Crossbite
  anterior, 126f
  elastic archwires for, 129
  elastics for, 129f–130f, 129–130
  posterior, 89, 89f, 125, 127
  quad helix for, 127–128, 128f
  rapid maxillary expansion for, 125–127, 126f–127f
  transpalatal arch for, 127, 128f
Cross-section stiffness, 34, 35t
Curve of Spee
  angulated, 110, 110f, 112f
  deep bite with, 109–110, 110f–111f
  reverse, 122, 122f
  stepped, 110, 110f
  use of straight or reverse-curved archwires, 110, 111f

D

Deep bite correction
  Class II, division 2 cases, 99–100, 100f–101f, 112–113
  continuous intrusion arch for, 105, 105f–106f
  curve of Spee correction, 109–110, 110f–111f
  differential mechanics, 111–112, 112f
  esthetics, 102f–103f, 102–103
  in normal- to low-angle patients, 101–102
  in transitional dentition cases
    description of, 113
    2 × 4 arch, 113f–115f, 113–115, 117f
  incisors
    lip relationship with, 102–103
    protrusion of, 100–101, 101f
    selective intrusion of, 103–107, 104f–107f, 107t, 114f
  microimplants for, 105–106, 107f
  overview of, 99–100, 100f
  selective molar extrusion for, 108, 108f–109f
  temporary anchorage devices for, 106, 107f
Dental open bite, 118
Differential mechanics, of space closure, 146–147, 151, 153
Distalization. See Canine distalization; Molar distalization.
Drawer movement, 94

E

Edgewise brackets, 76
Edgewise technique, 65, 66f
Elastic materials, 18
Elastic wires, 67, 67f, 129
Elastics
  Class II, 45–47, 46f–47f, 46t, 92f, 137, 152f, 153
  Class III, 45, 47
  coil springs vs, 48–49

crossbite, 129f–130f, 129–130
disadvantages of, 42
elastic behavior of, 18, 19f
elastomeric chains, 42–43
intraoral latex, 43–47, 44f–47f
ligatures, 49
natural rubber, 41
relaxation of, 42, 42f
synthetic polymers, 41–42
testing of, 43, 44f
Elastomeric chains, 42–43
Electrochemical corrosion, 27
Elgiloy, 28, 40
En masse protraction, 147, 147f
En masse retraction, 148f, 148–149
Equivalent force systems, 10, 10f
Extraction
  factors that affect, 145
  of molars, 112
  space closure after. See Space closure.
Extraoral appliances
  for anchorage control. See Anchorage control, extraoral appliances for.
  headgear. See Headgear.
  molar distalization using, 134–135

F

Facebow, 88–90, 134
Facemask, 153
Fatigue, 26
Force
  anteroposterior, 86
  as vector, 2f
  asymmetric application of, 91–95
  bracket-head screw used to apply, 130, 130f
  canine distalization, 27, 74
  constancy of, 4, 5f
  continuous, 4, 5f, 32
  definition of, 4
  distribution of, 4, 4f
  equivalent systems, 10, 10f
  friction affected by, 77
  in high-pull headgear, 86–87, 87f
  intermittent, 4, 5f
  interrupted, 4, 5f
  orthodontic use of, 17
  reverse headgear, 93
  space closure with maximum anchorage, 146, 147f
  transmissibility along its line of action, 7–8
Force elements
  cantilevers, 38f–40f, 38–39
  coil springs, 18, 39–41, 41f
  elastics. See Elastics.
  ideal type of, 27
  materials used in
    elastic behavior of, 18–19, 19f
    elastics. See Elastics.
    physical properties of, 17–27
    plastic, 18, 19f
    testing of, 24–27
    wires. See Wire(s).

Formability, 24
Friction
  bracket-wire. See Bracket-wire friction.
  during canine distalization, 73, 74f
  example of, 71–73
  force effects on, 77
  kinetic, 71
  static frictional force, 71
  tooth movement affected by, 73
Frictional systems, 71–80, 147–149, 148f–149f
Frictionless systems, 78f, 78–80, 150–153, 151f–152f
Frontomaxillary suture, 126

G

Gable bend, 60, 61f, 150
Gingival hyperplasia, 103
Gram-millimeters, 6
Gummy smile, 102f–103f

H

Headgear
  asymmetric, 92f
  cervical, 86, 87f, 108, 134, 143
  high-pull. See High-pull headgear.
  Kloehn, 143
  patient compliance with, 144
  reverse, 93f–94f, 93–95
  symmetric, 90, 90f
High-pull headgear
  anchorage control using, 86–87, 87f, 89f, 90
  transpalatal arch and, for skeletal open bite correction, 120–122, 122f
Hooke's law, 18
Hyrax screw, 125–126, 126f
Hysteresis, 29–30, 30f

I

Incisor(s). See also Mandibular incisors; Maxillary incisors.
  canine distalization effects on, 79, 79f, 148
  crown tipping of, 146
  extrusion of, 118–120, 120f
  lip relationship with, 102–103
  protrusion of, 100–101, 101f, 115, 117
  retraction of, 148
  selective intrusion of, 103–107, 104f–107f, 107t, 114f
  utility arch effects on, 115, 117
Inconsistent mechanics, 38f, 39
Inertia
  law of, 1
  moment of, 25, 25f
Interbracket distance
  friction and, 76–77
  in segmented arch technique, 78
  in stepped-arch mechanics, 63
  wire affected by, 20, 21f–22f, 32, 76–77
Intergranular corrosion, 26–27
Intermittent forces, 4, 5f

Interrupted forces, 4, 5f
Intra-arch appliances, for molar distalization, 144
Intramaxillary elastics, 43, 45f
Intraoral latex elastics, 43–47, 44f–47f

J
Jarabak ratio, 118
Jiggling effect, 69, 101

K
Kinetic friction, 71, 73
Kloehn headgear, 143

L
Laceback, 67–68, 68f, 142
Latex elastics, 43–47, 44f–47f
Law of acceleration, 1
Law of action and reaction, 1
Law of inertia, 1
Leveling
  description of, 35–36
  in deep bite correction with segmented arch mechanics, 105
  nickel titanium wires for, 37, 105
  with straight wire, 101
Ligatures, 49, 69, 77
Line of action
  adjustments to, 89
  transmissibility of force along, 7–8
Lip bumper, 85, 85f
Lip-incisor relationship, 102–103, 103f
Load/deflection rates of wire, 19–21, 20f–22f
Loop(s)
  deactivation of, 16
  description of, 27–28
  disadvantages of, 28
  moment-to-force ratio affected by configuration of, 13–16, 14f–15f
  objective of, 13, 27
  space closure uses of, 13–14, 150, 151f
Looped arch, 64f
Loss of alveolar support
  center of resistance affected by, 12, 13f
  moment-to-force ratios for teeth with, 12–13

M
Mandibular arch expansion, 130
Mandibular incisors. See also Incisor(s).
  crowding of, 94
  extrusion of, canine distalization as cause of, 79, 79f
  intrusion of, 107t
  protrusion of, 46–47, 48f
Martensite, 28
Martensitic transformation, 28
Material stiffness, 34
Maxillary arch constriction, 118
Maxillary expansion, rapid. See Rapid maxillary expansion.

Maxillary incisors. See also Incisor(s).
  overeruption of, 103, 103f
  selective intrusion of, 103, 107t
Maxillary molar distalization, 112
Maxillary retrusion, 94
Maximum anchorage, 145–146, 147f, 151
Mentalis muscle, 85
Mesially tipped molars, 139–143, 140f–142f
M/F ratio. See Moment-to-force ratio.
Microbiologic corrosion, 27
Microimplants
  anchorage control using, 95–96
  en masse retraction with, 148–149
  molar distalization using, 134f, 135
  molar intrusion with, for skeletal open bite correction, 122–123, 123f
  molar protraction using, 138, 139f
  molar uprighting uses of, 141, 142f
  selective incisor intrusion use of, 105–106, 107f
  zygomatic cortical bone, 123
Milled brackets, 76
Minimum anchorage, 145–146, 153
Moderate anchorage, 145–146, 151, 153
Modulus of elasticity, 18
Molar(s)
  buccal tipping of, 89f
  cervical headgear application on, 86, 87f
  deep bite correction by selective extrusion of, 108, 108f–109f
  distal tipping of, 88f, 116f
  eruption control of, using stainless steel wires, 122f
  extraction of, 112
  extraoral force application to, 86, 89
  extrusion of, 91, 108, 108f–109f, 141f, 151
  intrusion of, for skeletal open bite, 122–123, 123f
  mesially tipped, 139–143, 140f–142f
  moment-to-force ratio on, 146
  protraction of, 123, 123f, 138, 139f, 147
  rotation of, 143, 143f
  selective extrusion of, 108, 108f–109f
  tipback of, 110, 111f, 115, 116f, 135
  tipping of, 88f–89f, 89
  translation of, 87–88
  transpalatal arch for controlling vertical movements of
    description of, 89
    high-pull headgear and, 120–122, 122f
  uprighting of, 139–143, 140f–142f
Molar distalization
  extraoral appliances for, 134–135
  goals of, 133
  in transitional dentition, 133–134
  intra-arch appliances for, 144
  maxillary, 112
  microimplants for, 134f, 135
  Nance appliance for, 136, 137f
  open coil springs for, 135, 135f

  sliding jig for, 136–137, 137f
  summary of, 143–144
  superelastic wires for, 136, 136f
  temporary anchorage devices for, 135, 135f
  transpalatal arch with microimplant support for, 135, 135f
  2 × 4 arches for, 135–136
  vertical growth pattern considerations, 133–134
Molar offset, 64f, 65
Moment, 6, 7f
Moment of inertia, 25f, 25–26
Moment-to-force ratio
  center of rotation affected by changes in, 9f, 11
  description of, 6, 9
  formula for, 11
  interbracket distance and, 63
  loop configuration effects on, 13–16, 14f–15f
  loss of alveolar support effects on, 12–13
  of loops, 150
  on molars, 146
  T-loop, 15, 15f
Monocrystalline brackets, 50–51
Multistrand wires, 27, 28f, 31–32

N
Nance appliance, 83–84, 136, 137f
Natural rubber, 41
Newton's laws, 1, 8, 71, 145
Nickel titanium coil springs, 40–41, 41f
Nickel titanium wires
  clinical performance of, 30–31
  continuous intrusion arch and, 106f
  copper, 30
  discovery of, 28
  hysteresis of, 29–30, 30f
  leveling uses of, 37, 105
  metallurgic properties of, 28–29
  soft tissue irritation caused by, 76
  springback of, 23f
  stainless steel wires vs, 29
  stiffness of, 20–21, 22t, 36t
  superelasticity of, 29–30, 30f
  torsional properties of, 31, 32t
  working range of, 22t

O
Open bite
  dental, 118
  illustration of, 66f
  skeletal. See Skeletal open bite.
Oral environment, 26–27
Orthodontic appliances. See also Extraoral appliances.
  active units of, 18
  corrosive forces on, 26–27
  elements of, 17
  fatigue, 26

force elements used in. *See* Force elements.
oral environment effects on, 26–27
passive units of, 18
Orthopedic facemasks, 93. *See also* Reverse headgear.

P

Pitch angle, 41
Pitting corrosion, 26
Plastic brackets, 51
Plastic materials, 18
Posterior bite block, 119, 120f, 130, 140
Posterior crossbite, 89, 89f, 125, 127
Posterior teeth
  molars. *See* Molar(s).
  moment-to-force ratio for movement of, 151
  protraction of, 146–147
Power hooks, 77, 149f
Protraction
  en masse, 147, 147f
  of molars, 123, 123f, 138, 139f, 146–147
Protraction headgear. *See* Reverse headgear.
Protrusive smile, 102, 103f

Q

Q force, 91
Quad helix, 127–128, 128f

R

Rapid maxillary expansion
  crossbite treated with, 125–127, 126f–127f
  description of, 94
Reciprocal anchorage, 60
Rectangular wire, 31, 35t, 114
Relaxation, 42, 42f
Removable appliances, 11
Retraction of anterior teeth, 145
Reverse curve of Spee, 122, 122f
Reverse headgear, 93f–94f, 93–95
Reverse-curved wire
  anterior open bite correction using, 122, 122f
  deep bite with curve of Spee correction using, 110, 111f
Ricketts bioprogressive technique, 28
Rotation
  description of, 10
  of molars, 143, 143f
Round tripping, 101
Round wire, 28, 35t, 114
Rowboat effect, 35, 68–69, 142–143

S

Saliva, 76
Scissors bite, 129–130, 130f
Segmented arch technique
  anterior open bite correction using, 119
  deep bite correction with, 105
  interbracket distance in, 78, 78f
  space closure with, 150–153, 151f–152f
  utility arch vs, 117

Selective incisor intrusion, 103–107, 104f–107f, 107t, 114f
Selective molar extrusion, 108, 108f–109f
Shape-driven mechanics, 69
Shear, 18
Single rotation, 10
Skeletal open bite
  canting of maxillary occlusal plane associated with, 118, 119f
  diagnosis of, 118
  etiology of, 118
  treatment of
    goals for, 119
    incisor extrusion, 118–120, 120f
    molar intrusion with microimplant anchorage, 122–123, 123f
    posterior bite block, 119, 120f, 140
    transpalatal arch and high-pull headgear combination, 120–122, 122f
Sliding jig
  description of, 84
  molar distalization using, 136–137, 137f
Smile
  gummy, 102f–103f
  protrusive, 102, 103f
Space closure
  continuous archwire technique for, 150, 151f
  en masse retraction for, 148f, 148–149, 152f
  in frictional mechanics, 147–149, 148f–149f
  in frictionless mechanics, 150–153, 151f–152f
  loops for, 13–14, 150, 151f
  maximum anchorage for, 145–146, 147f, 151
  methods of, 145
  minimum anchorage for, 145–146, 153
  moderate anchorage for, 145–146, 151, 153
  segmented arch technique for, 150–153, 151f–152f
  two-stage, 148–149, 149f
Springback, 23, 23f
Springiness, 18
Springs. *See* Coil springs; Uprighting springs.
Stainless steel brackets, 49–50
Stainless steel coil springs, 40–41, 41f
Stainless steel wires
  elastic properties of, 32, 33t
  force degradation of, 49
  microbiologic corrosion of, 27
  molar eruption control using, 122f
  molar protraction uses of, 138
  multistrand, 27, 28f, 37
  nickel titanium wires vs, 29
  properties of, 27
  rounding the ends of, 138, 138f
  springback of, 23f
  stiffness of, 20–21, 22t, 31, 36t
  uprighting of molars using, 139
  working range of, 22t
Static equilibrium, 8, 8f

Static frictional force, 71, 73
Statically determinate force systems, 55–59
Step-down bends, 65
Stepped curve of Spee, 110, 110f
Stepped-arch mechanics, 63f–64f, 63–65
Step-up bends, 65, 69
Stiffness, 18
Straight arch, 64f
Straight wire
  deep bite correction using, 100
    with curve of Spee, 110, 111f
  mechanics of, 65–69, 142
Straight-pull headgear, 87, 88f
Strength of wire, 21–23, 22t
Stress, 17, 18f
Stress/strain rate, 18–19, 19f, 29
Subtraction of vectors, 3, 3f
Sum of vectors, 3, 3f
Superelastic wires, for molar distalization, 136, 136f
Superelasticity, 29–30, 30f
Synthetic polymers, 41–42

T

Teeth. *See also* Canine distalization; Incisor(s); Molar(s).
  anchorage value of, 83, 84f
  movement of. *See* Tooth movement.
Temporary anchorage devices
  deep bite correction using, 106, 107f
  description of, 95–96
  en masse retraction with, 148–149, 149f
  microimplants. *See* Microimplants.
  molar distalization using, 135, 135f
  molar intrusion with, for skeletal open bite correction, 122–123, 123f
  molar uprighting uses of, 141, 142f
  transpalatal arch and, 123f
Tension, 17
Three-point bending test, 24f, 24–25
Tipback angle, 114
Tipback of molars, 110, 111f, 115, 116f, 135
Tipping
  binding caused by, 74
  buccal, of posterior teeth, 126
  controlled, 8–9, 9f, 66, 146
  force distribution in, 4f
  of molars, 88f–89f
  uncontrolled, 8–9, 9f, 11, 13–14
  wire stiffness and, 74
T-loop, 14f–15f, 15, 151, 152f
Toe-in, 64f, 65
Tooth movement
  biomechanics of, factors that affect, 69
  controlled, 146
  energy sources of, in orthodontics, 17
  friction effects on, 73
  posterior, 151
  rotation, 10, 143, 143f
  tipping. *See* Tipping.
  translation, 10, 10f

Torque, 7
Toughness, 19
Transition temperature, 28, 30
Transitional dentition
  crossbite correction in
    quad helix for, 128
    rapid maxillary expansion for, 125, 126f
  deep bite correction in
    description of, 113
    2 × 4 arch, 113f–115f, 113–115, 117f
  molar distalization in, 133–134
Translation
  description of, 10, 10f
  of molars, 87–88
Transpalatal arch
  anchorage control using, 84–85, 89
  crossbite treated with, 127, 128f
  high-pull headgear and, for skeletal open bite correction, 120–122, 122f
  temporary anchorage device with, 123f
  with microimplant support, for molar distalization, 135, 135f
Transverse discrepancies, 125–130
Tweed anchorage bends, 64f, 65
Tweed technique, 83
2 × 4 arch
  deep bite correction using, 113f–115f, 113–115, 117f
  molar distalization using, 135–136
Two-tooth mechanics
  statically determinate force systems, 55–59
  stepped-arch mechanics, 63f–64f, 63–65
  straight-wire mechanics, 65–69
  V-bend arches, 60–63, 61f–62f

U

Uncontrolled tipping, 8–9, 9f, 11, 13–14
Uniform corrosion, 26
Uprighting of molars, 139–143, 140f–142f
Uprighting springs, 84, 84f
Utility arch
  advantages of, 115
  buccal bridge, 115, 116f
  incisor protrusion and intrusion caused by, 115, 117
  intraoral activation of, 116f
  molar anchorage reinforced with, 117
  molar tipback and extrusion caused by, 115
  sections of, 115, 116f
  segmented intrusion arch vs, 117

V

Variable modulus orthodontics, 34
Variable size orthodontics, 34
V-bend arches, 60–63, 61f–62f, 65f, 113–114, 150
Vectors, 2f–3f, 2–3
Vertical balancing, 153
Vertical discrepancies, 99–123
Vertical growth pattern
  molar distalization and, 133–134
  molar protraction and, 138
Vertical holding appliance, 144

W

Wire(s)
  β-titanium
    characteristics of, 31
    indications for, 37
    stiffness of, 20–21, 22t, 36t
    torsional properties of, 31, 32t
    working range of, 22t
  bracket and, frictional relationship between. See Bracket-wire friction.
  chrome-cobalt alloy, 28
  comparisons among, 31–32, 32t–33t
  cross-section stiffness of, 34, 35t
  elastic, 67, 67f, 129
  elastic behavior of, 18–19, 19f, 32t
  elasticity of, 18–20, 19f
  fatigue of, 26
  formability of, 24
  high-stiffness, 19–20
  interbracket distance effects on, 20, 21f–22f
  length of, 20
  load/deflection rates of, 19–21, 20f–22f
  loops. See Loop(s).
  low-stiffness, 20
  material stiffness of, 34
  moment of inertia, 25f, 25–26
  multistrand, 27, 28f, 31–32, 37t
  nickel titanium
    clinical performance of, 30–31
    continuous intrusion arch and, 106f
    copper, 30
    discovery of, 28
    hysteresis of, 29–30, 30f
    leveling uses of, 37
    metallurgic properties of, 28–29
    soft tissue irritation caused by, 76
    springback of, 23f
    stainless steel wires vs, 29
    stiffness of, 20–21, 22t, 36t
    superelasticity of, 29–30, 30f
    torsional properties of, 31, 32t
    working range of, 22t
  performance parameters of, 19–24, 20f–23f
  rectangular, 31, 35t, 114
  reverse-curved, 110, 111f, 122, 122f
  round, 28, 35t, 114
  second-order contact angles of, 21t
  selection of, 31
  shape of, 34–36
  size of, 20, 34–36
  springback of, 23, 23f
  stainless steel
    elastic properties of, 32, 33t
    force degradation of, 49
    microbiologic corrosion of, 27
    molar eruption control using, 122f
    molar protraction uses of, 138
    multistrand, 27, 28f, 37
    nickel-titanium wires vs, 29
    properties of, 27
    rounding the ends of, 138, 138f
    springback of, 23f
    stiffness of, 20–21, 22t, 31, 36t
    uprighting of molars using, 139
    working range of, 22t
  stiffness of, 19–21, 20f–22f, 22t, 31, 34–36, 36t–37t, 74
  straight. See Straight wire.
  strength of, 21–23, 22t
  stress application to, 18, 19f
  superelastic, for molar distalization, 136, 136f
  surface roughness of, 76
  testing of, 24f–25f, 24–27
  V-bend, 60
  working range of, 22t, 23, 25
Wire ligatures, 49, 77
Wire-bracket angle, 77–78
Working range, 22t, 23, 25

X

x-axis, 3–4

Y

y-axis, 4
Young's modulus, 18

Z

Zygomatic cortical bone microimplants, 123